METERS ᴬᴺᴰ SCOPES

HOW TO USE TEST EQUIPMENT

This book is dedicated to Susan Harris, my fellow writer, without whose help and inspiration this work would not have been possible.

METERS AND SCOPES

HOW TO USE
TEST EQUIPMENT

ROBERT J. TRAISTER

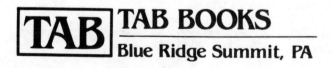

TAB BOOKS

Blue Ridge Summit, PA

7/93

15281184

FIRST EDITION
FOURTH PRINTING

Printed in the United States of America

Library of Congress Cataloging-in-Publication Data

Traister, Robert J.
    Meters and scopes.

    Includes index.
    1. Electronic measurements.   2. Electronic
instruments.   I. Title.
TK7878.T695   1988   621.3814′48   87-1966
ISBN 0-8306-0226-7
ISBN 0-8306-2826-6 (pbk.)

TAB BOOKS offers software for sale. For information and a catalog, please contact TAB software Department, Blue Ridge Summit, PA 17294-0850.

Questions regarding the content of this book should be addressed to:

    Reader Inquiry Branch
    TAB BOOKS
    Blue Ridge Summit, PA 17294-0214

Cover photograph courtesy of SENCORE, Inc., Electronic Test Equipment, 3200 Sencore Drive, Sioux Falls, SD 57107

Contents

Introduction

Anyone who has ever used, experimented with, or even dabbled in electronic equipment has found it necessary to perform a test or two and to measure certain electronic values. Often these measurements were conducted haphazardly and resulted in inaccurate readings due to inexperience in the use of electronic test instruments. Today's test instruments are user-oriented. While some may appear to be complex in nature and often are, most are relatively simple to operate and will provide accurate test measurements with the manipulation of a few controls. While their internal functions may be difficult to understand, their uses are not complex and almost anyone can learn to successfully operate these devices in a very short period of time.

This book is written as a primer for the beginner and as a reference source for the more experienced technician. The function and use of various test instruments is described, and practical information is provided that shows how to connect them to electronic circuits under test.

Today, most test instruments offer a myriad of test functions, performing many measurement operations with the convenience of a single device. Test equipment will differ from manufacturer to manufacturer, but once you know how to operate one brand, converting to another is usually only a matter of a few minutes familiarization.

By properly operating test equipment, the overall measurement process is made much more efficient. The informed user need not shy away from performing tests that were previously thought to be so complex that only a licensed engineer could attempt them.

Modern test instruments are designed for everyone. Those that were too complex for the average person to operate have gone the way of the dinosaur. It is hoped that this book will shed some light on the mystery that has unjustifiably surrounded electronic testing and will open a new world of experimentation, use, and repair to the average individual.

Chapter 1

Matter and Measurement

It is impossible to pinpoint the precise moment in history when man stopped wandering about the earth in search of food and shelter and began reflecting upon the forces that governed his existence. His first feeble attempts to understand his environment became the basis for science.

If there are any roots to Western science, they no doubt lie under the rubble that was once ancient Greece. The Greeks reasoned that the earth was a sphere, that substances must have basic particles, and even proposed a crude atomic theory. Unfortunately, they were not prone to self-criticism and did not practice a system of rigid experimentation. Man may never know how many universal truths were postulated by these early Greeks, for as time passed into the Middle Ages, most information of a scientific nature was lost due to neglect and ignorance.

The millennium between 486 A.D. and the end of the 15th century is commonly referred to as the Dark Ages. They were dark because the brilliance of human reason fell prey to the superstition and mysticism that ran rampant throughout Europe at that time. The fact that science was kept alive at all was mainly due to the ancient craft known as alchemy. The alchemist spent most of his time searching for some magic process whereby he could turn base metals into gold. Although these searches were fruitless, he did, through recorded data, provide some useful information and very fertile ground for the inquisitive men we now call scientists. If the ties of

1

superstition are ever completely severed, credit for the initial incision would no doubt be given to two early giants of modern science, Galileo and Newton. It was with them that the scientific revolution began.

Before going into a discussion of measurement, it is appropriate to understand a little about matter and mass, two commonly used terms in science. Matter is most often defined as anything that has mass and occupies space. It is relatively easy to form a mental image of some object as it occupies space. The meaning of mass, however, may not be quite so obvious. Mass is best defined as the property of matter that gives it inertia or opposition to a change in state of motion or rest.

The terms weight and mass can be used synonymously on earth, especially at sea level. The difference between mass and weight is that the mass of an object is the same anywhere in the universe, but its weight is not. This difference is apparent to our astronauts as they travel through outer space. A man possessing a definite amount of mass might weigh 150 pounds at the surface of the earth but would be weightless in an orbiting satellite. This fact illustrates that a body of matter can have mass without weight, but weight always indicates the presence of mass. Weight is actually a measure of the gravitational force pulling a mass toward the center of the earth.

MEASUREMENT AND ITS IMPORTANCE

Practically every field of science deals in some way with either the structure or the measurement of matter. If a modern scientist were asked to select what he considered to be the most important single factor in the rapid advance of present-day civilization, he might well choose the ability to measure quantities accurately. Without realizing it, practically every act of our daily lives involves measurement of some kind. In driving to work or school, our speed is measured in miles per hour. Upon arriving at our destination, we determine whether or not we are late by looking at a clock, the instrument of time measurement. How handicapped the chemist would be if he could not measure the myriad quantities of chemicals in his experiments! Picture the chaotic results of a carpenter's efforts if he could not measure the length of the timbers used in constructing a home. So important is the ability to measure that one of Britain's great scientists, Lord Kelvin, believed that unless a person can describe the topic of his study with measurements, he knows nothing about that topic. It can be seen that measurement plays a large part in our everyday lives. Measurement is also the very foundation of the study of electricity and electronics.

PRINCIPLES OF MEASUREMENT

Basically, all measurements, regardless of the type, are accomplished in the same manner. To make a measurement, you must compare the dimensions of the quantity to be measured with the dimensions of a known, standard quantity. The quantity to be measured is then said to be so many times larger or smaller than the known, standard quantity.

Any measurement can be divided into two parts. The first part tells how many times larger or smaller the unknown quantity is, and the second part tells the standard or reference used for comparison. For example, assume that a half-back runs the length of a football field in ten seconds. From this measurement, his running time is determined to be ten times greater than the standard or reference, or one second. This comparison of the halfback's time to the time of one second tells how long it takes him to run the length of the field.

To make measurements meaningful to other people, a reference must be used which has one exact meaning to all who use it. In the above example, everyone concerned with the measurement must agree on the length of time of one second. Otherwise, ten seconds would indicate a different amount of time to each person.

The British realized the need for exact and standard references. According to legend, one of England's many monarchs attempted to develop a system of standardization in which he, the king, was the standard. A common unit of measure, the foot, was developed by this king. His own foot was the physical standard of reference; but unfortunately, this standard was made unavailable upon the death of the king. Another standard developed by this man was the rod. A rod equalled the length of a line composed of twenty lords who stood chest to back. But again, unfortunately, the lean years shrunk the rod, while the bountiful years stretched it and thus the "standard" took on different meanings at different times. Beyond these difficulties, however, this king had great foresight, for his idea of standard units is still in use today. Though many systems of measurement have been devised, only the British system and the metric system have received popular following in this country.

THE METRIC SYSTEM

The metric system is used by most European countries and by scientist throughout the world. Any system used for the measurement of matter is ultimately based on three basic dimensions: length, mass, and time. Scientists use a form of the metric system known as the meter-kilogram-second (mks) system, in which the unit of length is the meter, the unit of mass is the kilogram, and the unit of time is the second.

The meter was originally defined as one ten millionth of the distance along a meridian extending from the North Pole through Paris to the Equator. At the time of this definition, a physical model was created and housed at the International Bureau of Weights and Measures in Sevres, France. The physical standard is made of a platinum-iridium alloy bar upon which two marks are etched. The distance between the two marks when the bar is at the temperature of melting ice is one meter. Modern investigation shows that this physical model varies slightly from time to time and is therefore not truly accurate. At present, however, this bar remains the physical standard, although in the near future, scientists may adopt a new and unvarying standard based upon the length of one wave of a specific kind of light energy.

The kilogram uses as its standard a block of platinum-iridium alloy called the International Prototype Kilogram, also preserved at Sevres, France. The kilogram is based on the gram (1/1000 of a kilogram), originally defined as the mass of one cubic centimeter of pure water measured at four degrees centigrade. This temperature was chosen because near four degrees centigrade, the density of water is practically independent of temperature.

The second is defined as 1/86,400 of a mean solar day, the mean solar day being the average time required for the earth to make one rotation on its axis while revolving about the sun. In recent years, it has been found that this standard also has irregularities, and in the near future, an atomic clock based on the vibration of a cesium atom may be adopted.

A subdivision of the metric system is known as the centimeter-gram-second (cgs) system, in which submultiples of the meter and kilogram are used. A centimeter is one hundredth of a meter; the gram and second have already been explained.

In many situations, the quantities to be measured may be extremely large or extremely small, and the basic units may prove too cumbersome for ease of manipulation. For example, the length of our galaxy, the Milky Way, is approximately 946,800,000,000,000,000,000 meters. It should also be kept in mind that compared to other astronomical distances, the length of the Milky Way is small. Thus, to the astronomer, the unit meter is not

Prefix	Abbr.	Power of Ten	Value
Tera	T	10^{12}	Million million
Giga	G	10^9	Thousand million
Mega	M	10^6	Million
Kilo	K	10^3	Thousand
Hecto	h	10^2	Hundred
Deca	dk	10^1	Ten
-	-	10^0	One
Deci	d	10^{-1}	Tenths
Centi	c	10^{-2}	Hundredths
Milli	m	10^{-3}	Thousandths
Micro	μ	10^{-6}	Millionths
Nano	n	10^{-9}	Thousand Millionths
Pico	p	10^{-12}	Million Millionths
Femto	f	10^{-15}	-
Atto	a	10^{-18}	-

Fig. 1-1. Commonly used prefixes with their meanings.

convenient. A much more meaningful description of this distance would be 100,000 light years, where the light year is defined as the distance light travels in one year.

In the metric system, a prefix is attached to one of the basic units to provide a unit more consistent with the dimensions of the quantity involved. A list of the commonly used prefixes and their meanings is given in Fig. 1-1. A common example of the use of prefixes is the microsecond, which is one-millionth of a second.

THE BRITISH GRAVITATIONAL SYSTEM

A second system, the one common to this country and Great Britain, is known as the foot-pound-second system. This is the system with which the average person in this country is most familiar. The fps system differs from the metric system, in that the pound is not a unit of mass as is the kilogram, but is a unit of force. The unit of mass in the fps system is the *slug*, which is equal to approximately 14.6 kilograms. Formerly, standards were maintained for the fps system; but now, all these units are defined in terms of metric standards. The relationship between metric and British units are shown in Fig. 1-2.

THE MEASUREMENT OF LIGHT

The study of light has provided man with a fascinating but most perplexing problem in science. Man has learned to generate, control,

Metric to British	British to Metric
1 km. = 0.62137 mi.	1 mi. = 1.6093 km.
1 m. = 3.2808 ft.	1 ft. = 0.3048 m.
1 cm. = 0.3937 in.	1 in. = 2.5400 cm.
1 kg. = 2.2046 lb.	1 lb. = 0.4536 kg.
1 gm. = 0.0353 oz.	1 oz. = 28.3490 gm.

Note: The metric mass units are equated to the force in pounds that they would exert at sea level on earth.

Fig. 1-2. Comparison of metric and British units.

and measure light energy very effectively. Through measurements of the faint light coming from distant stars and planets, man has learned practically all that is now known about the objects in outer space. Unfortunately, the exact structure of light is still a mystery. It is well known that light is a form of energy, but the physical form in which this energy exists is not known. Nevertheless, two theories have been advanced concerning the nature of light.

One of these theories proposes the existence of light as tiny packets of energy called *photons*. Photons can contain various quantities of energy, the amount being dependent upon the color of the light involved. Photons of light at the blue-violet end of the spectrum contain more energy than photons of red light.

The second theory pictures light rays as consisting of electromagnetic waves of very short length. There is strong evidence to indicate that light may exist in either of the above states, depending on the conditions under which it is observed. If this is true, there is no single model which can be constructed to illustrate the dual nature of light.

One of the more important measurements associated with light energy is that of wavelength measurement. Light may be analyzed by assuming it consists of waves similar to the ripples which are generated when a ball is dropped into a pool of water, as shown in Fig. 1-3. The waves that are generated consist of a number of cycles, such as the one shown between points A and B. In traveling from A to B, the wave has gone through all of its possible variations and therefore has completed an entire cycle of events. In traveling from B to C, the wave would simply repeat the variations that occurred between A and B. The number of these complete cycles per second is called the *frequency* of the wave. If the wave illustrated

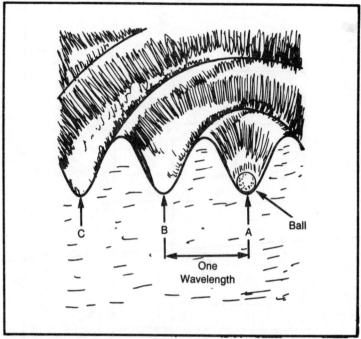

Fig. 1-3. Light can be assumed to consist of waves similar to the ripples created when a ball is dropped into water.

completes one cycle in one-twentieth of a second, it would have a frequency of twenty cycles per second.

The distance between a point on wave A and a corresponding point on an adjacent wave B is called the wavelength of the wave. Wavelength may be measured in any of the distance units described previously, such as inches, feet, meters, or centimeters, etc. Light waves have extremely short wavelengths of less than one-millionth of an inch. Figure 1-4 shows the wavelength in centimeters (cm) of various types of electromagnetic waves, including light waves.

THE MEASUREMENT OF HEAT

Heat is a form of radiant energy which is very similar in nature to light. For many purposes, heat energy (often called thermal energy) may be considered to be light energy of a wavelength too long for detection by the human eye. Heat energy, like light energy, is often thought to be present in all materials in small massless packets called *phonons* (also called quanta). The number of *phonons* present in a material is directly related to temperature. Heat energy

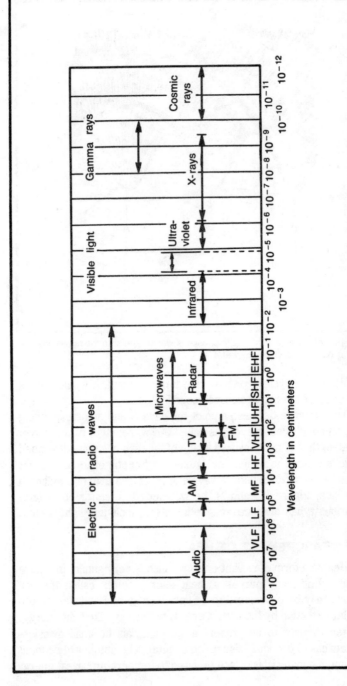

Fig. 1-4. Electromagnetic wavelengths in centimeters.

8

can be obtained as a result of chemical reactions, nuclear reactions, friction, and electrical energy.

The quantity of heat a substance contains is normally measured in one of the following basic units: a calorie, which is the quantity of heat necessary to raise the temperature of one gram of water one degree Centigrade; or the British Thermal Unit, which is the quantity of heat necessary to raise the temperature of one pound of water one degree Fahrenheit.

From these definitions, it may be determined that the temperature of an object is by no means a measure of the amount of heat it contains. A small quantity of water in a test tube can be raised a degree in temperature by the heat from a wooden match. In contrast, consider the great amount of heat required to raise the temperature of the water in a swimming pool one degree. It is evident that these two quantities of water, while they may be at the same temperature, contain vastly different quantities of heat. Temperature indicates the extent to which a body has been heated rather than the amount of heat which it contains. Although many scales have been designed for the measurement of temperature, only the Celsius (more commonly called Centigrade scale) and the Fahrenheit scales will be discussed in this text. The main difference between these two scales, which are illustrated in Fig. 1-5, is the values arbitrarily

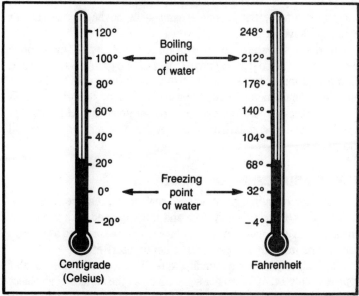

Fig. 1-5. The relationship of the Centigrade and Fahrenheit scales.

assigned to the freezing and boiling points of water. On the Centigrade scale, the freezing and boiling temperatures are 0 degrees and 100 degrees, respectively; while on the Fahrenheit scale, they are 32 degrees and 212 degrees, respectively. Equations for conversion from Fahrenheit to Centigrade or from Centigrade to Fahrenheit can be found in most basic physics and chemistry books.

THE MEASUREMENT OF FORCE

Although the word force usually causes a person to form a mental image of some type of mechanical system, the origin of many forces is electrical in nature. The operation of the electrical motors in vacuum cleaners, refrigerators, and other home appliances is entirely dependent upon the interaction of electrical forces. Because force occupies such a prominent position in the theory of electrical devices, a knowledge of the units used to measure force is necessary.

Specifically, force is defined as that quantity which causes acceleration (change in motion) of a material body. The pound, a unit with which we are all familiar, is the unit of force in the British system. A pound is the amount of force necessary to impart an acceleration of one foot per second to a mass of one slug.

In the mks system, the force unit is the *newton* (nt) and is defined as that force that will give a mass of one kilogram an acceleration of one meter per second per second. Thus, a newton would be the approximate force felt on one's hand while holding a one-quarter pound package of butter.

In the cgs system, the unit of force is the *dyne*. One dyne is the amount of force that will give a mass of one gram an acceleration of one centimeter per second per second. A dime coin held on the fingertips would exert a downward force of about 2450 dynes. This example indicates that the dyne is a rather small unit of force. Figure 1-6 shows a comparison between the newton, the dyne, and the pound.

THE MEASUREMENT OF PRESSURE

Another quantity which acts on matter is pressure. *Pressure* is defined as force per unit area and is expressed as a force unit divided by an area unit. Thus, pressure may be expressed as pounds per square inch, dynes per square centimeter, etc.

The atmosphere surrounding the earth exerts a pressure of approximately 14.7 pounds per square inch on the surface of an object placed at sea level. This normal sea level pressure is sometimes

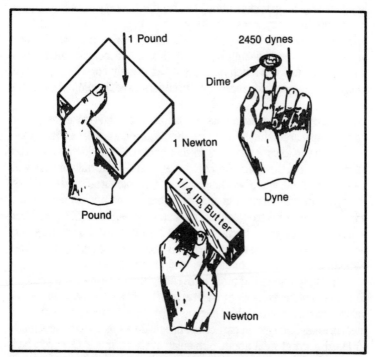

Fig. 1-6. A comparison of the newton, the dyne, and the pound.

used as a unit of pressure called an *atmosphere* (atm). One atmosphere, therefore, is a unit of pressure equal to approximately 14.7 pounds per square inch. Pressure may also be measured in *bars* or *microbars*. One bar is equal to 14.5 pounds per square inch or one million dynes per square centimeter.

Pressure may also be measured by comparing it to the pressure exerted by a column of mercury. Thus, a pressure may be stated as 7 millimeters (mm) of mercury, indicating a pressure equal to that exerted by a column of mercury 7 centimeters high. Similarly, a pressure of 10 centimeters (cm) of mercury would be the pressure exerted by a column of mercury 10 centimeters high. Over the years, many different units have been devised for the measurement of pressure. However, those mentioned here are the most common.

SUMMARY

In dealing with electronic circuits, you will constantly have need of some form or unit of measurement in order to describe the many values of the components that make up the device or circuit. This is a fact which certainly cannot be disputed. Measurement plays a large role in every person's daily life in both obvious and not so obvious ways. The amount of study and experimentation that has gone into the establishing of standard units of measurements has made all of our lives much easier and has greatly contributed to the many technological advances that scientists have developed over the years.

Electronic measurements are sometimes thought of as being very complicated simply because the required knowledge is not familiar. The following chapters in this book will provide some basic information on the common components and their values that will be encountered when working with electronic circuits, as well as the types of measurements and tests that can be performed in both a maintenance and troubleshooting capacity on these circuits. This is an important step, because before accurate measurements can be taken, it is necessary to learn as much as possible about the various components, conditions, and standard units.

Probably one of the most interesting things that the newcomer to the field of measurement will learn is that it is possible to measure values that are not tangible, such as heat, light, frequency, etc. The measurement of light, in particular, has made it possible for scientists to learn a great deal about our solar system, much of which had remained a mystery for many thousands of years.

Through the study of measurement and a basic knowledge of the many instruments available to perform these checks on circuits, the reader will learn much about the operation of electrical and electronic devices and should be able to make any repairs necessary.

Chapter 2
Electronic Values and Components

In order to proceed with the measuring of devices, components, and whole systems, it is absolutely essential to have a basic understanding of electronic terms and values, as well as the quantities and conditions which they describe. Of course, most readers are familiar with the more common electronic terms, such as voltage, amperage, resistance, etc. But do you really understand what these terms mean and how they are applied?

This chapter will provide some basic electrical and electronic theory in a simple and easily understandable manner. The information will provide you with the necessary foundation in the electronic functioning of devices in order that you will be able to understand the results obtained from the many types of measurements that will be discussed in later chapters in this book.

VOLTAGE

In every system in which a transformation of energy occurs, there must be a primary agent which supplies the initial energy to the system. In a system made up of wires and components, such as an electric circuit, this primary agent is commonly referred to as the *source* for the circuit. Flashlight cells and automobile batteries are good examples of electrical sources. Although a device that supplies electrical energy to a circuit is called the source, it must be remembered that the source does not *create* energy; rather, it is merely a device in which some other form of energy is transformed into electrical energy. In the case of a battery, for example, chemical energy is converted into electrical energy.

It is common for persons to refer to voltage as something which is "flowing in a circuit." This is not really accurate. Voltage does not flow; it is a force which activates a flow of current in a circuit. (Current will be discussed in detail later in this chapter.) Voltage is a force and can be likened to a bat which strikes a baseball. The movement of the ball can be compared with the flow of current, but the bat is the force which causes the movement of the ball.

The force which causes this movement can also be referred to as potential energy. This is the energy contained by an object due to its position. For example, a hammer and the earth form a system of masses in which exchanges of energy can take place by means of the earth's gravitational field. Assume that the hammer is suspended by a string in a position one meter above a nail. As a result of gravitational attraction, the hammer will experience a force pulling it downward toward the center of the earth. If the string is suddenly cut, the force of gravity will pull the hammer downward against the nail, driving it into the wood. While the hammer is suspended above the nail, it has the ability to do work because of its elevated position in the earth's gravitational field. Since energy is the ability to do work, the hammer contains energy, or potential energy. The amount of potential energy available is equal to the product of the force required to elevate the object and the height to which it is elevated.

In most cases, however, the total potential energy in a system is of little practical importance. A good example of this is an old-fashioned cuckoo clock. Figure 2-1 shows such a clock in which the gear mechanism used to turn the hands is operated by a slowly falling weight. In Fig. 2-2, a metal weight is suspended on a chain beneath the clock so that the gear will be rotated as the weight pulls downward on the chain. With a given length of chain, the weight can only fall a short distance to position x, its lowest possible position. To begin, assume that the weight having a mass of one kilogram is at position x. If the weight is then raised against the force of gravity from position x to position y, a distance of one meter, work will have been performed on the weight. The force required to overcome gravity is 9.8 newtons for a mass of one kilogram. The work accomplished can be calculated by the use of the equation shown in Fig. 2-3. Thus, the work accomplished in raising the weight from position x to position y is found to be 9.8 joules.

Fig. 2-1. A gear mechanism in this clock is used to turn the hands by means of a weight.

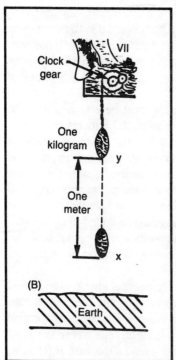

Fig. 2-2. The weight is suspended on a chain.

15

```
Given:     F  =  9.8 newtons
           d  =  1.0 meter

Solution:  W  =  Fd
           W  =  9.8 newtons × 1.0 meter
           W  =  9.8 joules
```

Fig. 2-3. This equation can be used to determine work accomplished in the clock example.

Close examination of this example shows that the weight now has capabilities which it did not have when it was in position x. If the weight is allowed to fall from y to x, work will be accomplished by the weight as it falls. Neglecting friction, the amount of work recovered when the weight falls is 9.8 joules, exactly equal to the work expended in raising the weight.

It is no mere coincidence that the work required to raise the weight (9.8 joules) and the amount of work the raised weight is able to perform are equal. This equality stems from one of the fundamental laws of physics that states that energy can be neither created nor destroyed but can be transformed from one kind to another.

The potential energy gained by a body is equal to the work done on that body in moving it from its original position to its final position. Because 9.8 joules of work was done on the clock weight in raising it from position x to position y, the weight gained 9.8 joules of potential energy.

In the case of the clock weight, it must be stressed that the weight in position y did not contain a total of 9.8 joules of energy. It *gained* 9.8 joules of energy in addition to the energy it contained in position x. For the weight to have zero potential energy, it would have to be placed at the earth's center of gravity. Because the weight could not possibly fall all the way to the earth's center, the total potential energy of the weight could never be fully utilized. Of far greater interest is the additional energy that is added to the system as the weight is raised from its lowest position to its highest position, since this represents the amount of energy which can be recovered as work.

To simplify problems dealing with potential energy, the lowest position of a body is used as a reference point and the body is considered to have zero potential when in this position. This simplification is similar to the system used in measuring altitude where sea level rather than the center of the earth is used as the zero reference.

The potential energy of the upper position of the body could then be computed with respect to the lower position. The result would be the difference of potential, or the potential difference, between the two positions and a true measure of the work that may be recovered. The volt is the basic unit which is used to express potential difference.

It can be seen, then, that in most electronic circuits, only the difference of potential between two points is of importance and the absolute potentials of the points are of little concern. Very often, it is convenient to use one standard reference for all of the various potentials throughout a piece of equipment. For this reason, the potentials at various points in a circuit are generally measured with respect to the metal chassis on which all parts of the circuit are mounted. The chassis is considered to be at zero potential and all other potentials are either positive or negative with respect to the chassis. When used as the reference point, the chassis is said to be at *ground potential*.

Occasionally, rather large values of voltage may be encountered, in which case the volt becomes too small a unit for convenience. In a situation of this nature, the kilovolt (kV), meaning 1,000 volts, is frequently used. For example, 20,000 volts would be written as 20 kV. In other cases, the volt may be too large a unit, as when dealing with very small voltages. For this purpose, the millivolt (mV), meaning one thousandth of a volt, and the microvolt (μV), meaning one millionth of a volt are used. For example, .001 volt would be written 1 mV, and .000025 volt would be written as 25 μV.

In the everyday language of electronics, the number of volts between two points is expressed in several different ways, including voltage, potential, potential difference, and electromotive force, or EMF. Strictly speaking, each of these terms indicates a specific quantity. However, they are quite frequently used interchangeably. EMF, for example, should only be used when referring to the force which causes charges to move through a source of voltage.

To be a practical source of voltage, the potential difference in a circuit must not be allowed to dissipate, but must continuously be maintained. As one electron leaves the concentration of negative charge, another must be immediately provided to take its place or the charge will eventually diminish to the point where no further work can be accomplished. A voltage source, therefore, is a device which is capable of supplying and maintaining voltage while some type of electrical apparatus is connected to its terminals. The internal action of the source is such that electrons are continuously removed from one terminal, which becomes positive and also simultaneously supplied to the second terminal, which assumes a negative charge.

Electromotive force is that force which moves charges within a voltage source. This name is misleading, however, because electromotive force cannot be measured in the conventional force units of newtons or pounds but is measured in volts. Thus, potential difference and electromotive force (EMF) are both measured in volts even though they are slightly different quantities and have different values in a given source.

There are two basic types of voltage, ac volts and dc volts. Dc voltage is most often supplied by batteries, while ac voltage is that which is derived from the standard outlets in a home. The terms ac and dc are abbreviations for alternating current and direct current. Current and voltage are closely tied to one another, so it is difficult to describe one value without including the other.

When speaking of voltage, it is necessary to know the type of voltage under discussion. In basic electrical circuits, voltage is measured between two points, the positive pole and the negative pole. Contacts from both of these poles must be applied to the circuit to be powered or to the measuring device in order to establish a flow of current. When properly connected to an electronic circuit, or *load*, a power source will cause current to flow from the negative pole to the positive one.

Dc voltage circuits have fixed poles. The polarity remains constant, in that one contact will always be positive while the other will be negative. Ac power supplies maintain a continual reversal of polarity. During one-half of the ac cycle, one pole will be positive while the other is negative. During the next half of the cycle, the pole that was formerly positive switches to negative while the former negative pole becomes positive. The polarity of the supply undergoes a complete reversal. The rate of polarity change, or alternation, in ac circuits is measured in *hertz* (Hz). Most of us are familiar with alternating current. Our household power is of this type and reverses at a rate of 60 Hz, which means that the polarity reverses itself 60 times in one second. This rate of change is called the *ac frequency*.

CURRENT

All electric and electronic devices utilize electron movement. This is known as current. In the early years of electrical study, electric current was erroneously assumed to be a movement of positive charges from positive to negative. This assumption is a concept that became entrenched in the minds of many scientists.

Since it has been proven that electrons (negative charges) move through a wire, electron current will be used throughout this explanation of electric current. The direction of electron movement, as mentioned in the previous discussion on voltage, is from a region of negative potential to a region of less negative potential or more positive potential. Therefore, electric current can be said to flow from a negative potential to a positive potential. The direction is determined by the polarity of the voltage source. Thus, it can be seen why current and voltage are so closely intertwined in a discussion of electricity.

Electric current is generally classified into two general types, direct current and alternating current. A direct current flows continuously in the same direction, whereas an alternating current periodically reverses direction.

Electric current, then, is defined as the directed movement of electrons. *Directed drift*, therefore, is another term which can be used interchangeably with current, since this is defined as the migration of electrons from one end of a conductor to another due to potential difference. This is particularly helpful in differentiating between the random and directed motion of electrons. However, current flow is the term most commonly used in indicating a directed movement of electrons.

The magnitude of current flow is directly related to the amount of energy that passes through a conductor as a result of drift action. An increase in the number of energy carriers or an increase in the energy of the electrons would provide an increase in current flow. When an electric potential is impressed across a conductor, there is an increase in the velocity of the electrons, thus causing an increase in the energy of the carrier. There is also the generation of an increased number of electrons providing added carriers of energy. The additional number of free electrons is relatively small; hence, the magnitude of current flow is primarily dependent upon the velocity of the existing electrons.

The magnitude of current flow is affected by the difference of potential and the condition of the crystal lattice. Current flow is affected by the difference of potential in the following manner. Initially, the mobile electrons are given additional energy because of the repelling and attracting electrostatic field. These electrons in turn collide with atoms, releasing this energy to the atom. The electron that strikes the atom may deliver all or part of its energy to the atom. Assuming that it delivers only part of its energy, as indicated in Fig. 2-4, the electron will continue to travel under the

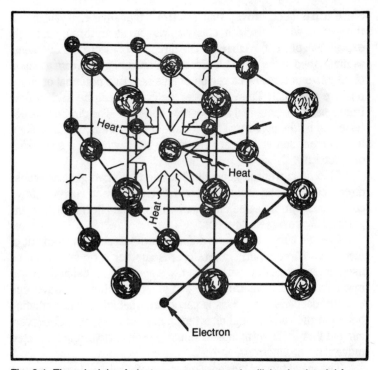

Fig. 2-4. The principle of electron movement and collision in pictorial form.

influence of the electric field. If the potential difference is increased, the electric field will be stronger, the amount of energy imparted to a mobile electron will be greater, and the current will be increased. If the potential difference is decreased, the strength of the field is reduced, the energy supplied to the electron is diminished, and the current is decreased.

The magnitude of current is measured in *amperes*. A current of one ampere is said to flow when one coulomb of charge passes a point in one second. Expressed as an equation:

$$I = \frac{Q}{T}$$

where: I = current in amperes
         Q = charge in coulombs
         T = time in seconds

Frequently, the ampere is much too large a unit. Therefore, the milliampere (ma), which is one thousandth of an ampere, or the microampere (μa), which is one-millionth of an ampere are used. The device to measure current is called an *ammeter*.

RESISTANCE AND CONDUCTANCE

We know that the directed movement of electrons constitutes a current flow. We also know that the electrons do not move freely through a conductor's crystalline structure. Some materials offer little opposition to current flow, while others greatly oppose current flow. This opposition to current flow is known as *resistance*, and the unit of measure is the *ohm*. The standard of measure for one ohm is the resistance provided at zero degrees centigrade by a column of mercury having a cross-sectional area of one square millimeter and a length of 106.3 centimeters. A conductor has one ohm of resistance when an applied potential of one volt produces a current of one ampere. The symbol used to represent the ohm is the Greek letter omega, Ω.

Resistance, although it is an electrical property, is determined by the physical structure of a material. The resistance of a material is governed by many of the same factors that control current flow.

The magnitude of resistance is determined in part by the number of mobile electrons available within the material. Because a decrease in the number of mobile electrons will decrease current flow, it can be said that the opposition to current flow (resistance) is greater in a material with fewer free electrons. Thus, the resistance of a material is determined by the number of free electrons available in a material.

Resistance is also dependent upon the length of the *mean free path* of the electrons. This is defined as the average distance between collisions of the electrons and atoms. As the mean free path of an electron is shortened, the current flow decreases and the resistance increases. Likewise, as the mean free path of an electron is lengthened, current flow increases and resistance decreases.

Specific resistance is a standard that has been developed to compare the resistance value of conductors. Specific resistance, or *resistivity*, is the resistance in ohms offered by the unit volume of a substance. In the English system, the unit volume of a material is the volume of a cylindrical conductor which is one foot long and one mil in diameter. This is illustrated in Fig. 2-5. Each different conducting material has its own specific resistance, which is determined by a number of factors. Figure 2-6 lists the specific resistances of some common materials. Of the materials shown, silver has the least resistance and is thus the best conductor.

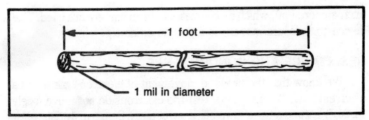

Fig. 2-5. Unit volume is the volume of a cylindrical conductor measuring one foot in length and one mil in diameter.

The term which is just the opposite of resistance is *conductance*. Conductance is the ability of a material to pass electrons. The factors that affect the magnitude of resistance are exactly the same for conductance, but they affect conductance in the opposite manner. Therefore, conductance is directly proportional to area, and inversely proportional to the length and specific resistance of the material. The temperature of the material will also be a factor; but assuming a constant temperature, the conductance of a material can be calculated if its specific resistance is known.

The unit of conductance is the *mho*, which is ohm spelled backwards. Whereas the symbol used to represent the magnitude of resistance is the Greek letter omega(Ω), the symbol used to represent conductance is ℧. The relationship that exists between resistance and conductance is a reciprocal one. The reciprocal of a number is one divided by that number. If the resistance of a materi-

	Specific resistance at 20°C	
	---	---
Substance	Centimeter cube (microhms)	Circular mil-foot (ohms)
Silver	1.629	9.8
Copper (drawn)	1.724	10.37
Gold	2.44	14.7
Aluminum	2.828	17.02
Carbon (amorphous.)	3.8 to 4.1	. . . .
Tungsten	5.51	33.2
Brass	7.0	42.1
Steel (soft)	15.9	95.8
Nichrome	109.0	660.0

Fig. 2-6. Chart showing specific resistance of common materials.

al is known, dividing its value into one will give its conductance. Also, if the conductance is known, dividing its value into one will give its resistance. The value of conductance is usually very small. Since the possibility of error is great when working with numbers of minute magnitudes, the value can also be expressed in micromhos. Therefore, the conductance of 0.000772 mhos would be 772 micromhos.

POWER

Power is the rate of doing work per unit of time. Work results from a force acting on a mass over a distance. The operation of electrical circuits involves a force (voltage) acting on a mass (electrons) over a distance. The amount of time required to perform a given amount of work will determine the power expended. Expressed as an equation, the relationship between power, work, and time is:

$$P = \frac{W}{t}$$

where: P = power in watts
W = work in joules
t = time in seconds

Since energy is the capacity to do work, power can also be defined as the time rate of developing or expending energy. In every electrical circuit, electrical energy is transformed into heat energy. Thus, by measuring the amount of heat energy given off by an electrical circuit in a given amount of time, the amount of electrical power consumed in a circuit can be determined.

The unit of measure for electrical power is the *watt*. The amount of wattage used in a circuit can be determined by multiplying the supply voltage by the supply current, as expressed in amperes. Assume a circuit with an input voltage of 9 volts and a current of 0.5 ampere. The power consumption of this circuit would be (9 × 0.5 = 4.5), or 4.5 watts.

Electrical components are often given a *power rating*. The power rating in watts indicates the rate at which the device converts electrical energy into another form of energy such as light, heat, or motion. An example of such a rating would be the comparison of a 150-watt lamp to a 100-watt lamp. The higher wattage rating of the 150-watt lamp indicates that it is capable of converting more electrical energy into light energy than the lamp with the lower rating. Other common examples of devices rated in this manner are soldering irons and small electric motors.

In some electrical devices, the wattage rating indicates the maximum power the device is designed to dissipate rather than its normal operating power. For example, a 150-watt lamp dissipates 150 watts when operated at the rated voltage printed on the bulb. In contrast, a device such as a resistor is not normally given a voltage or a current rating. A resistor is given a power rating in watts and can be operated at any combination of voltage and current as long as the power rating is not exceeded. In most circuits, the actual power dissipated by a resistor will be considerably less than the resistor's power rating. In well-designed circuits, a safety factor of 100% or more is allowed between the actual dissipation of the resistor in the circuit and the power rating listed by the manufacturer. The wattage rating of the resistor, then, is the maximum power the resistor can dissipate without damage from overheating. Power is measured in watts with a wattmeter.

CAPACITANCE

Capacitance is defined as the property of an electrical device or circuit that tends to oppose a change in voltage. Capacitance is also a measure of the ability of two conducting surfaces separated by some form of nonconductor to store an electric charge. The device used in electrical circuits to store a charge by virtue of an electrostatic field is called a *capacitor*. The larger the capacitor, the larger the charge that can be stored. Two conductors separated by a nonconductor, then, exhibit the property which is called capacitance, because this combination can store an electric charge. In order to understand capacitance, it is necessary to become familiar with some of the theories and laws of electrostatics.

When a charged body is brought into close proximity with another charged body, there is a force that causes the bodies to attract or repel one another. If the charged bodies possess the same sign of charge, a repelling force will exist between the two bodies. If they have unlike signs, there will be a force of attraction between them. The force of attraction or repulsion is caused by the electrostatic field that surrounds every charged body. If a material is charged positively, it has a deficiency of electrons. If it is charged negatively, it has an excess of electrons. The direction of the electrostatic field is represented by lines of force perpendicular to the charged surface and shown originating from the positively charged material. Each

24

line of force is drawn in the form of an arrow and is shown pointing from positive to negative.

The force between charges is described by Coulomb's law: "The force existing between two charged bodies is directly proportional to the product of the charges and inversely proportional to the square of the distance separating them."

If a test charge is inserted in an existing electrostatic field, it will move toward one or the other of the charged areas which is causing the field to exist. The direction of movement will depend on whether the test charge is positive or negative. A positive test charge placed in a field moves in the direction that the line of force points, or from positive toward negative. In this case, the test charge will be an electron, and since the electron is negative, it will move in a direction opposite to that of the positive charge. In other words, an electron in an electrostatic field will move against the arrow from negative toward positive. This action is illustrated in Fig. 2-7.

If Coulomb's law is analyzed in connection with this figure, it can be seen that the greater the distance between the electron and the positive charge, the less the force of attraction.

One important characteristic of electrostatic lines of force is that they have the ability to pass through any known material. The simplest type of capacitor consists of two metal plates separated by air. It has been established that a free electron inserted in an electrostatic field will move. The same is true, with qualifications, if the electron is in a bound state. The material between the two

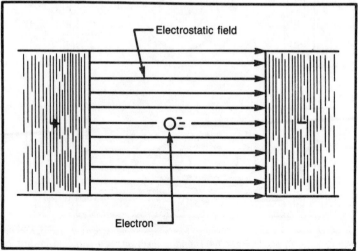

Fig. 2-7. Electron movement in an electrostatic field.

charged surfaces of Fig. 2-7 (air in this example) is composed of atoms containing bound orbital electrons. Since the electrons are bound, they cannot travel to the positively charged surface. Therefore, the resultant effect will be a distortion of the electron orbits. The bound electrons will be attracted toward the positive surface and repelled from the negative surface. This effect is illustrated in Fig. 2-8. In part A, there is no difference in charge placed across the plates and the structure of the atom's orbits is undisturbed. If there is a difference in charge across the plates as shown in part B, the orbits will be elongated in the direction of the positive charge.

Because energy is required to distort the orbits, energy is transferred from the electrostatic field to the electrons of each atom between the charged plates. Since energy cannot be destroyed, the energy required to distort the orbits can be recovered when the electron orbits are permitted to return to their normal positions. This effect is analogous to the storage of energy in a stretched spring. A capacitor can thus store electrical energy. An illustration of a simple capacitor and its schematic symbol is shown in Fig. 2-9. The conductors that form the capacitor are called *plates*. The material between the plates is called the *dielectric*. In part B of this figure,

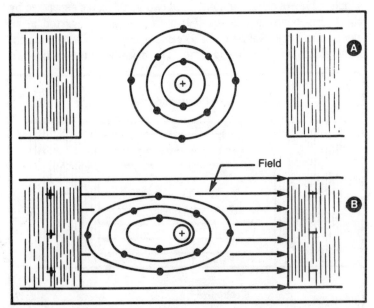

Fig. 2-8. In an electrostatic field, bound electrons are attracted toward the positive surface and repelled from the negative surface.

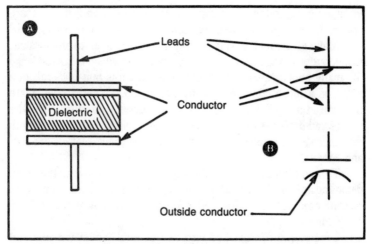

Fig. 2-9. A simple capacitor shown in pictorial and schematic form.

the two vertical lines represent the connecting leads. The two horizontal lines represent the capacitor plates. Notice that the schematic symbol in B and the simple capacitor diagram in A are similar in appearance. In a practical capacitor, the parallel plates may be constructed in various configurations (circular, rectangular, etc.); but the cross-sectional area of the capacitor plates is tremendously large in comparison to the cross-sectional area of the connecting conductor. This means that there is an abundance of free electrons available in each plate of the capacitor. If the cross-sectional area and the plate material of the capacitor are the same, the number of free electrons in each plate must be approximately the same. It should be noted that there is a possibility of the difference in charge becoming so large as to cause ionization of the insulating material to occur (cause bound electrons to be freed). This places a limit on the amount of charge that can be stored in the capacitor.

Capacitance is measured in a unit called the *farad*. This unit is attributed to Michael Faraday, a scientist who performed many early experiments with electrostatics and magnetism. It was discovered that for a given value of capacitance, the ratio of charge deposited on the plate to the voltage producing the movement of charge is a constant value. This constant value is a measure of the amount of capacitance present. The symbol used to designate a capacitor is C. The capacitance is equal to one farad when a voltage changing at the rate of one volt per second causes a charging current of one amp to flow. The farad can also be defined in terms of charge and voltage. A capacitor has a capacitance of one farad if it will store one coulomb of charge when connected across a potential of one volt.

Many electronic circuits require capacitors of quite small value. Consequently, the farad is a cumbersome unit which is far too large for many applications. The microfarad, which is one-millionth of a farad, is a more convenient unit. The symbols used to designate microfarad are μF and MFD. In high frequency circuits, even the microfarad becomes too large, and the micromicrofarad (one-millionth of a microfarad) is used. The symbols for micromicrofarads are $\mu\mu$F and MMFD.

To avoid confusion and the use of double prefixes, the term picofarad (pF) is preferred in place of micromicrofarad. In powers of ten, one picofarad, (or one micromicrofarad) is equal to 1×10^{-12} farad.

Capacitors are classified into two general types: variable and fixed. Variable capacitors are constructed in a fashion that allows the value of capacitance to be varied over a prescribed range. Fixed capacitors are categorized by the type of dielectric used. Capacitors are also manufactured in various sizes and shapes. Some are small tubular resistor-like devices, and others are molded, rectangular, flat components. Figure 2-10 shows some of the more common shapes of capacitors.

Although the value of a capacitor may be indicated by printing on the body of the device, many capacitive values are indicated by the use of a color code. The colors used to represent the numerical value of a capacitor are the same as those used to identify resistance values. In this type of system, a series of colored dots (sometimes bands) is used to denote the value of the capacitor. These dots will contain information about the capacitor, such as working voltage, temperature coefficient, etc.

It is already known that capacitance is the ability to oppose a change in applied voltage. When the applied voltage is changed, the capacitor charges or discharges until the voltage on the capacitor is equal to the new value of applied voltage. At the time when the capacitor voltage is equal to the source voltage, no more current flows. Because a capacitor reacts to a voltage change by producing a counter electromotive force (CEMF), a capacitor is said to be *reactive*. The opposition of a capacitor is therefore called *reactance*, and is measured in ohms. The opposition offered by a capacitor to alternating current is termed *capacitive reactance* and is designated by X_C.

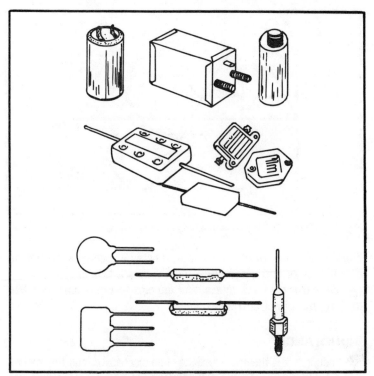

Fig. 2-10. Capacitors may take many different physical forms.

Although no current actually flows through a capacitor, circuit current will exist whenever a capacitor charges or discharges. If a capacitor is connected across an alternating voltage source, an alternating current will flow as the capacitor tries to charge and discharge in step with the voltage. If a sine wave of voltage is applied to a capacitor, a sine wave of current will result. Since current in a capacitive circuit is maximum when the rate of change of voltage is maximum, the current waveform will be offset 90 degrees from the voltage waveform. This is shown in Fig. 2-11. Notice that when the voltage is passing through zero (maximum rate of change), the current is maximum. When the voltage is at its peak value (minimum rate of change), the current is zero. Thus, in a capacitive circuit, the current leads the voltage by 90 degrees. This phase relationship is shown in Fig. 2-12.

Capacitors are used for a large number of purposes in electronic circuits. They have the ability to pass alternating current while

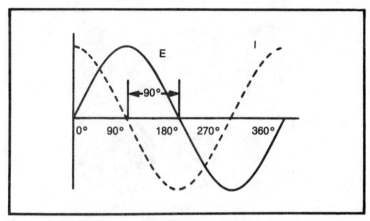

Fig. 2-11. Current/voltage relationships in a capacitive circuit.

blocking the flow of direct current. This is a highly useful trait in circuits where both alternating and direct currents are present. A capacitor can pass the alternating current to other circuits while keeping the direct current confined.

INDUCTANCE

Inductance is the characteristic of an electrical circuit that makes itself evident by opposing the starting, stopping, or changing of current flow. Anyone who had ever had to push a heavy load (wheelbarrow, car, etc.) is aware that it takes more work to start the load moving than it does to keep it moving. This is because the load possesses the property of *inertia*. Inertia is the characteristic of mass which opposes a change in velocity. Therefore, inertia can hinder in some ways and help in others. Inductance exhibits the same effect on current in an electric circuit as inertia does on velocity of

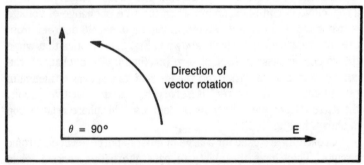

Fig. 2-12. In a capacitive circuit, current leads the voltage by 90 degrees.

a mechanical object. The effects of inductance are sometimes desirable and sometimes undesirable.

There is another way to explain inductance. If the current in an electric circuit increases, a self-induced voltage (counter electromotive force) opposes this change and delays the increases. If the current decreases, a self-induced voltage tends to aid or prolong the current flow. Thus, the most noticeable effect of inductance in a circuit is that the current can neither increase nor decrease as fast as it can in a circuit without inductance. All connectors and coils induce within themselves an EMF which will oppose any change in the internal current. Therefore, any conductor or coil, by definition, possesses the property of inductance. In order for a coil or piece of wire to possess this property, current need not be present. From this, it can be seen that a wire hanging in free space will still possess the property of self-induction or inductance.

The unit for measurement of inductance is the *henry*, which is so-named in honor of Joseph Henry, who discovered the property of self-induction of a coil. A coil is said to possess an inductance of one henry if an EMF of one volt is induced in the coil when the current through the coil is changing at the rate of one ampere per second.

Figure 2-13 shows examples of two types of inductors and their schematic symbols. The air-core type inductor is most frequently used in circuits that are above the audio range. The iron-core type is widely used in the audio range. The iron-core type is usually made of laminated sheets of iron to reduce core losses.

In order to understand inductance further, it is necessary to understand the factors which affect computing the total inductance in a circuit or circuits. When two coils are positioned so that flux lines from one cut the turns of the other, they are said to exhibit *mutual inductance*. The action of mutual inductance in producing a EMF is basically the same as that of self-inductance, the main difference being that self-inductance is the property of a single coil, while mutual inductance is the property of two or more coils acting together. Mutual inductance is also measured in henrys and is designated by the letter, M.

The induced EMF of an inductor is dependent on the mutual inductance. Mutual inductance, in turn, is dependent on the physical dimensions of the two coils, the number of turns on each coil, the permeability of the cores, and the *coefficient of coupling*. Coefficient of coupling is dependent upon the distance between two coils and on the position of the coil axis with respect to each other. In other

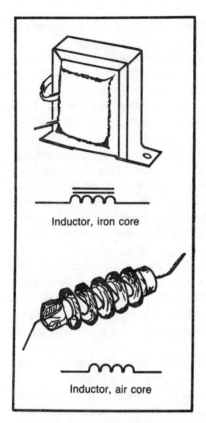

Fig. 2-13. Examples of two types of inductors.

Inductor, iron core

Inductor, air core

words, coefficient of coupling is a measure of how much of the flux from one coil cuts the turns of the other coil.

In Fig. 2-14, the inductors are seen to be very close. In fact, almost all of the flux from L1 cuts all the turns of L2. Under these conditions, the coefficient of coupling is maximum or very near unity. Under ideal conditions, all the flux from one coil cuts all the turns of the other coil and the coefficient of coupling is unity or one. This

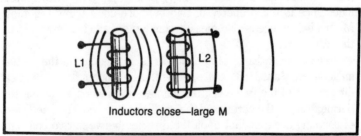

Inductors close—large M

Fig. 2-14. An example of inductive coupling.

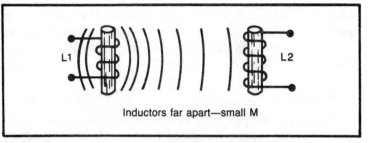

Inductors far apart—small M

Fig. 2-15. When inductors are separated by increasing distances, the flux field decreases rapidly in direct relation to the distance of separation.

is a condition which may be approached but never reached in practical applications.

In Fig. 2-15, the inductors are seen to be physically separated by a considerable distance. Since the strength of a flux field decreases rapidly with distance, it is easy to see how very few of the flux coils from coil L1 will cut the turns of L2. Therefore, the coefficient of coupling is small, making the mutual inductance also small.

In Fig. 2-16, the relative position of the axes of the two coils is seen to be perpendicular. Under these conditions, there are virtually no conductors positioned at right angles to the flux lines. If none of the flux of L1 is able to cut the turns of L2, the coefficient of coupling is zero and the mutual inductance is zero.

The opposition that an inductance offers to a changing current, then, can be referred to as self-induced voltage or counter electromotive force (CEMF), which is measured in volts. However, opposition to current flow is normally measured in ohms rather than volts. Because a coil reacts to a current changing by generating a CEMF, a coil is also said to be reactive, and the opposition of a coil is therefore termed reactance. Since more than one type of

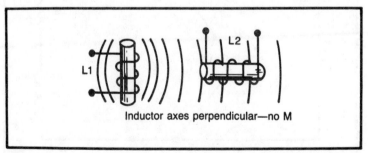

Inductor axes perpendicular—no M

Fig. 2-16. The relative position of the axes of two coils can be used to control coupling.

reactance exists, such as capacitive reactance, the subscript L is added to denote inductive reactance. Thus, the opposition offered by a coil to alternating current will be termed *inductive reactance* and is designated by X_L.

IMPEDANCE

Impedance is a term used to describe a property which is not quite as specific as the terms that have already been discussed. Impedance can be descriptive of any electrical circuit or device which serves to impede the flow of current. To see this, refer to Fig. 2-17, which shows a circuit composed of a coil and a resistor connected in series and placed across an ac source. This circuit displays both resistive and reactive characteristics. Because of this, the total opposition of the circuit cannot be called resistance or reactance, because it contains some of each. The total opposition, then, is called *impedance* and is represented by the letter Z. Like resistance and reactance, impedance is measured in ohms.

RESISTORS

Now that an understanding has been obtained of the more common electronic values, it is also necessary to discuss the components that exhibit these values. Although resistance is a property of every electrical component, at times its effects will be undesirable. However, resistance is used in many varied ways. Resistors are components which are manufactured to possess specific values of resistance. They are manufactured in many different types

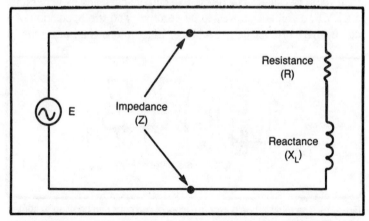

Fig. 2-17. An ac circuit composed of an inductor and a resistance.

Fig. 2-18. Schematic representation of a fixed resistor.

and sizes. When drawn in a schematic, a resistor is represented by a series of jagged lines, as shown in Fig. 2-18.

One of the most common types of resistors is the carbon resistor shown in Fig. 2-19. This type, with the leads extending parallel to the length of the resistor, is known as an *axial lead resistor. Carbon resistors* have as their principal ingredient the element carbon, as their name implies. The specific resistance of carbon lies between the range of 20 to 27 ohms per cir-mil-ft. Carbon has a conductance that is approximately one five-hundredth that of silver. In the manufacture of carbon resistors, fillers or binders are added to the carbon to obtain various resistor values. Examples of filters are clay, Bakelite, rubber, and talc. These fillers are doping agents and cause the overall conduction characteristics to change. Carbon resistors are the most common type found because they are easy to manufacture, inexpensive, and have a tolerance that is adequate for most electronic applications. Their prime disadvantage is that they have a tendency to change value as they age. One other disadvantage of carbon resistors is their limited power-handling capability.

The disadvantages of carbon resistors can be overcome by the use of *wire-wound resistors*. A wire-wound resistor is shown in Fig. 2-20. This type of resistor has a very accurate value and possesses a higher current-handling capability than the carbon resistor. The material frequently used to manufacture wire-wound resistors is German silver, which is composed of copper, nickel, and zinc. The quantities of these elements present determine the resistivity, but the resistivity will be primarily dependent on the percentage of nickel present. One disadvantage of wire-wound resistors is that it takes

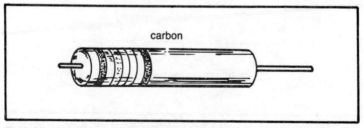

carbon

Fig. 2-19. A carbon resistor is a physically small device and is often color-coded with different identifying bands.

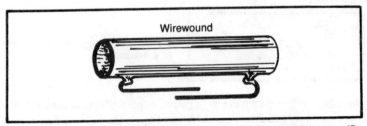

Fig. 2-20. A wire-wound resistor uses many turns of wire to establish a specific resistance.

a large amount of wire to manufacture a resistor of high ohmic value, thereby increasing the cost.

Resistors are also classified as either *fixed* or *variable*. Those already discussed are classed as fixed. Fixed resistors will have one value and will not be subject to any changes other than temperature, age, etc. Figure 2-21 shows another type of fixed resistor, the *tapped resistor*. The tapped resistor has several fixed taps and furnishes more than one resistance value. Another type, the *sliding contact resistor* (Fig. 2-22) has an adjustable collar that can be moved to tap off any resistance value within the range of the resistor. Such a resistor does not completely fulfill the requirements of a fixed resistor. However, the collar is normally adjusted to a desired position and kept there. The sliding contact resistor thus serves as a fixed tapped resistor.

There are two types of variable resistors, the *potentiometer* and the *rheostat*. These are pictured in Fig. 2-23. There is a slight difference between these two types. Rheostats have two

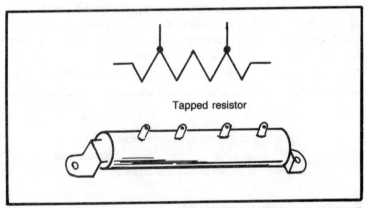

Fig. 2-21. The tapped resistor offers several different resistance ranges, all housed within a single component body.

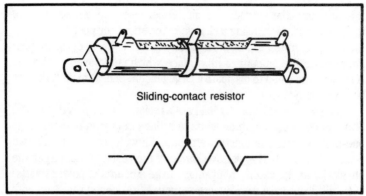

Fig. 2-22. An adjustable resistor uses a sliding metal collar which intercepts the resistance windings to vary the overall resistive value.

connections, one fixed and one movable. The potentiometer has three contacts, two fixed and one variable. Generally, the rheostat has a limited range of values and a high current-handling capability. The potentiometer has a wide range of values, but it has a limited current-handling capacity.

When a current is passed through a resistor, heat is developed within the component. The resistor must be capable of dissipating

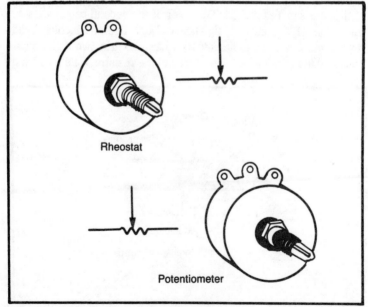

Fig. 2-23. Pictorial and schematic drawings of a rheostat and a potentiometer.

this heat into the surrounding air. Otherwise, the temperature of the resistor rises, causing a change in resistance or possibly causing the component to burn out. The ability of a resistor to dissipate heat depends on the amount of its surface which is exposed to the air. A resistor designed to dissipate a large amount of heat must therefore be large, physically.

The heat dissipation capability of a resistor is measured in watts. Some of the more common wattage ratings of carbon resistors are: one-eighth watt, one-fourth watt, one-half watt, one watt, and two watts. The higher the wattage rating of the resistor, the larger the physical size. Resistors that require large amounts of power (watts) to be dissipated are usually wire-wound. Wire-wound resistors with wattage ratings up to 50 watts are not uncommon.

Value Designations of Resistors

There are two systems used to designate the ohmic value of resistors. One is the standard system and the other is called the body-end-dot system. The standard system is the one that is most frequently used. Both systems make use of a color code. The position of the color and the color code indicate the value of the resistor.

In the standard system, four bands are painted on the resistor, as shown in Fig. 2-24. The color of the first band indicates the value of the first significant digit. The color of the second band indicates the value of the second significant digit. The third color band represents a decimal multiplier by which the first two digits must be multiplied to obtain the resistance of the resistor. The final band

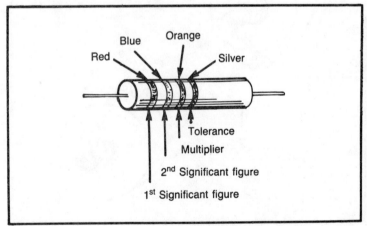

Fig. 2-24. Each resistor color band designates a different numerical value.

Color	Significant Figure	Decimal Multiplier	Resistance Tolerance
Black	0	1	Percent ±
Brown	1	10	—
Red	2	100	—
Orange	3	1,000	—
Yellow	4	10,000	—
Green	5	100,000	—
Blue	6	1,000,000	—
Violet	7	10,000,000	—
Gray	8	100,000,000	—
White	9	1,000,000,000	—
Gold	—	.1	5
Silver	—	.01	10
No Color	—		20

Fig. 2-25. The resistor color-code chart.

indicates the tolerance of the resistor. The colors used for the bands and their corresponding values are shown in Fig. 2-25.

Using the example in Fig. 2-24, since red is the color of the first band, the first significant digit is 2. The second band is blue; therefore, the second significant figure is 6. The third band is orange, which indicates that the number formed as a result of reading the first two bands is multiplied by 1000. In this case, 26 × 1000 = 26,000 ohms. The last band indicates the tolerance. Its color is silver, and thus, the tolerance is 10 percent. The allowable limit of variation in ohmic values will be from 23,400 to 28,600 ohms.

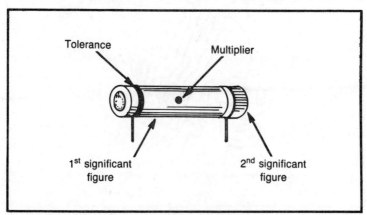

Fig. 2-26. The body-end-dot system of color coding.

The body-end-dot system of color coding is used with radial-lead resistors. The colors correspond to the same numbers as given in Fig. 2-25. The colors, positions, and their significance are shown in Fig. 2-26.

CAPACITORS

The capacitor is the device which is used in electrical circuits to store a charge. As mentioned earlier in this chapter, capacitors can be divided into two classifications: fixed and variable. This discussion will explain the different types of both fixed and variable capacitors in greater detail.

Variable capacitors, which allow capacitance to be varied over a specific range, can be classified as either the *rotor-stator* type or the *trimmer* type. The rotor-stator types use air as the dielectric, which is the insulating material used between the capacitor plates. The amount of capacitance is varied by changing the position of the rotor plates (movable plates). This changes the effective plate area of the capacitor. When the rotor plates are fully meshed between the stator plates, the capacitance is at maximum. The rotor-stator type of capacitor is illustrated in Fig. 2-27. This type of variable capacitor finds wide application in such devices as table model radios.

The trimmer type of variable capacitor consists of two plates which are separated by a dielectric other than air. The capacitance is varied by changing the distance between the plates. This is

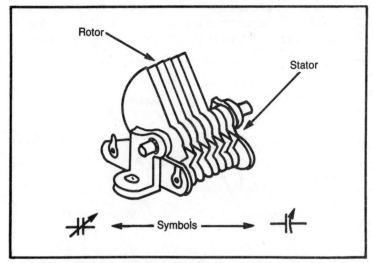

Fig. 2-27. A variable capacitor.

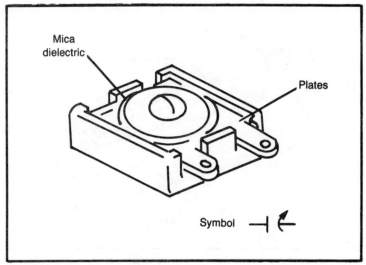

Fig. 2-28. A trimmer capacitor.

ordinarily accomplished by means of a screw which forces the plates closer together. The trimmer capacitor is shown in Fig. 2-28.

Fixed capacitors are categorized by the type of dielectric used. There is more variety with regard to fixed capacitors because of this. Each will be discussed individually.

Paper Capacitors

A *paper capacitor*, as its name implies, is one which uses paper as its dielectric. The construction of a typical paper capacitor is shown in Fig. 2-29. It consists of flat, thin strips of metal-foil conductors, which are separated by the dielectric material, in this case, paper. In this type of capacitor, the dielectric used is waxed paper.

Paper capacitors usually range in value from approximately 4 picofarads to 300 microfarads. Normally, the voltage limit across the plates will rarely exceed 600 volts. Paper capacitors are sealed with wax to prevent the harmful effects of moisture.

Mica Capacitors

Mica capacitors consist of alternate layers of mica and plate material. Their capacitance is very small, usually in the picofarad range. Although small physically, the mica capacitors have a high voltage-handling capacity. Figure 2-30 shows a cutaway view of a mica capacitor.

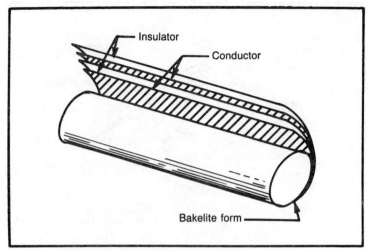

Fig. 2-29. Basic construction of a paper capacitor.

Oil Capacitors

Oil capacitors are often used in radio transmitters where high output power is desired. Oil-filled capacitors are nothing more than paper capacitors which are immersed in oil. The oil-impregnated paper has a high dielectric constant, which lends itself well to the production of capacitors that have a high value. Many capacitors will use oil with another dielectric material to prevent arcing between the plates. If an arc should occur between the plates of an oil-filled

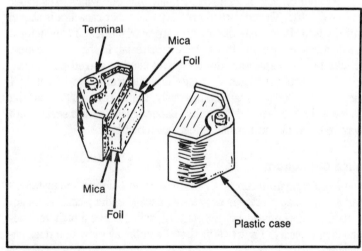

Fig. 2-30. Construction of a mica capacitor.

capacitor, the oil will tend to reseal the hole caused by the arc. These types are often referred to as self-healing capacitors.

Ceramic Capacitors

Ceramic capacitors are so-named because of the use of ceramic as the dielectric. One type of ceramic capacitor uses a hollow ceramic cylinder as both the form on which to construct the capacitor and as the dielectric material also. The plates consist of thin films of metal deposited on the ceramic cylinder.

A second type of ceramic capacitor is manufactured in the shape of a disk. After leads are attached to each side of the capacitor, the device is completely covered with an insulating, moisture-proof coating. Ceramic capacitors usually range from one picofarad to 0.01 microfarad and may be used with voltages as high as 30,000 volts.

Electrolytic Capacitors

Electrolytic capacitors are used where a large amount of capacitance is required. As the name implies, electrolytic capacitors contain an electrolyte. This electrolyte can be in the form of either a liquid (wet electrolyte) or a paste (dry electrolyte). Wet electrolytic capacitors are no longer in popular use today due to the care needed to prevent spilling of the electrolyte.

Dry electrolytic capacitors consist essentially of two metal plates between which is placed the electrolyte. In most cases, the capacitor is housed in a cylindrical aluminum container which acts as the negative terminal of the capacitor, as shown in Fig. 2-31. The positive

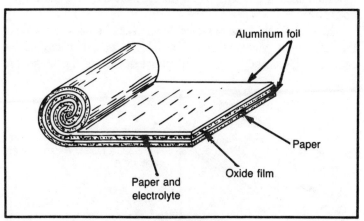

Fig. 2-31. Construction of an electrolytic capacitor.

terminal (or terminals if the capacitor is of the multisection type) is in the form of a lug on the bottom end of the container. The size and voltage rating of the capacitor is generally printed on the side of the aluminum case.

Internally, the electrolytic capacitor is similar in construction to the paper capacitor. The positive plate consists of aluminum foil covered with an extremely thin film of oxide which is formed by an electrochemical process. This thin oxide film acts as the dielectric. Next to and in contact with the oxide, a strip of paper or gauze which has been impregnated with a paste-like electrolyte is placed. The electrolyte acts as the negative plate of the capacitor. A second strip of aluminum foil is then placed against the electrolyte to provide electrical contact to the negative electrode (electrolyte). When the three layers are in place, they are rolled up in a cylinder as shown in Fig. 2-32.

Electrolytic capacitors have two primary disadvantages. First of all, they are polarized, and second, they have a low leakage resistance. What this means is that if the positive plate were accidentally connected to the negative terminal of the source, the thin, oxide-film dielectric will dissolve and the capacitor will become a conductor (i.e., it will short). The polarity of the terminals is normally marked on the case of the capacitor. Because electrolytic capacitors are polarity sensitive, their use is ordinarily restricted to dc circuits or circuits where a small ac voltage is superimposed on a dc voltage. Special electrolytic capacitors are available for certain ac applications, such as motor-starting capacitors. Dry electrolytic capacitors vary in size from about four microfarads to several thousand microfarads and have a voltage limit of approximately 500 volts.

In general, the type of dielectric used and its thickness govern the amount of voltage that can be safely applied to any of the capacitors discussed here. If the voltage applied is high enough to

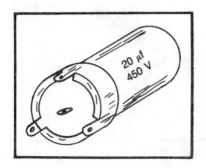

Fig. 2-32. A typical electrolytic capacitor.

cause the atoms of the dielectric material to become ionized, an arc-over will take place between the plates. If the capacitor is not self-healing, its effectiveness will be impaired. The maximum safe voltage of a capacitor is called its *working voltage* and is indicated on the body of the capacitor. The working voltage is determined by the type and thickness of the dielectric. If the thickness of the dielectric is increased, the distance between the plates is also increased, and thus, the working voltage will be increased. Any change in the distance between the plates will cause a change in the capacitance of the device. Because of the possibility of voltage surges (brief high-amplitude pulses), a margin of safety should be allowed between the circuit voltage and the working voltage of a capacitor. The working voltage should always be higher than the maximum circuit voltage.

SEMICONDUCTOR DEVICES

Semiconductors are those devices which are made up of materials with a conductivity halfway between that of a conductor and an insulator. These devices are made from crystalline materials which are chemically treated with impurities and then combined in sandwich layer style. The finished crystalline chip is then mounted in a small metal or plastic case through which leads protrude for attachment to the internal structure. Semiconductors contain no moving parts and are not subject to damage from vibration or moderate shocks. Semiconductor devices include many different types of components that perform a myriad of electronic functions. Each will be discussed individually.

TRANSISTORS

Semiconductors that can amplify a signal are called *transistors*. A transistor utilizes a small change in current to produce a large change in voltage, current, or power. The transistor, then, may function as an amplifier or as an electronic switch. There are many different types of transistors with individual characteristics, but the theory of operation is basic to all of them.

The advent of the transistor has opened a completely new field for the development of portable equipment. The compactness and ruggedness of transistorized equipment has allowed the manufacture of portable equipment that was previously impractical. Transistors are now being used in mobile equipment, test equipment, recording equipment, photographic equipment, hearing aids, radios, etc. In other words, transistors may be used in almost any application where low- and medium-power electron tubes are used.

Figure 2-33 shows a drawing of one type of transistor, the bipolar transistor. There are three leads protruding from this device which are connected to the emitter, collector, and the base of the component. Transistors have the ability to indirectly increase electrical power, and this is often rated in terms of multiplication. Shown in Fig. 2-34 are the two basic types of bipolar transistors, which are labeled pnp and npn. Pnp transistors are composed of a negative semiconductor material (n) sandwiched between two sections of positive material (p). The reverse is true of the npn units, which sandwich the p material between two sections of n material.

The various terms for describing certain electronic conditions were discussed earlier. Similar terms are used to describe conditions and ratings of semiconductors or solid-state devices. Most solid-state devices are described in two ways: limiting conditions and characteristics. Additionally, the manufacturer will usually give each device a manufacturer's type number and a packaging type. This refers to the case the component is mounted in.

Certain symbols will be used as abbreviations for describing characteristics and operating parameters of these devices. While there are many such abbreviations, only the most prevalent ones will be discussed here.

The first abbreviation stands for *device dissipation*, and is abbreviated P_t. This rating is most often expressed in watts, although in certain devices it may be expressed in milliwatts. P_t describes the maximum amount of power in watts that can be

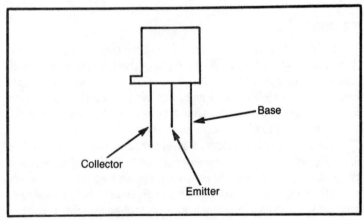

Fig. 2-33. A drawing of a typical bipolar transistor.

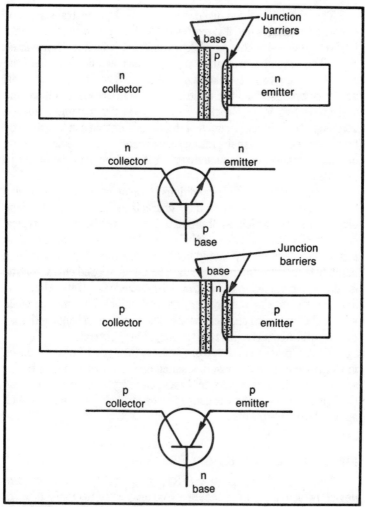

Fig. 2-34. Schematic drawings and construction of the two basic types of bipolar transistors.

dissipated by the component and still remain within the manufacturer's ratings.

Another important rating is that of *collector current*, abbreviated I_c. I_c describes the largest amount of current which can be drawn through the collector of the transistor without exceeding manufacturer's ratings. This condition is normally described in amperes of dc current, although it may be described in milliamperes for lower power devices.

The third condition used to describe bipolar transistors is *breakdown voltage*, and it is expressed in three different forms. First is the collector-to-base breakdown voltage, which is abbreviated V_{CVO}. The second abbreviation is V_{CEO}, which is a description of the collector-to-emitter breakdown voltage. The third and final abbreviation describes the emitter-to-base breakdown voltage and is displayed as V_{EBO}. Breakdown voltage is the maximum applied voltage to the various portions of the transistor that can be tolerated by the device. When these ratings are exceeded, the device is being operated outside of manufacturer's ratings and may soon fail to operate.

In addition to limiting conditions, bipolar transistors may also be described as to operating characteristics. One of these characteristics, which is abbreviated h_{FE}, stands for the typical current gain of the device. This is also known as the *amplification factor*. This figure is not stated in any specific term but uses a single number to indicate the multiplication factor. A second characteristic is the *typical gain bandwidth*, abbreviated f_T. Generally, this indicates the maximum frequency the transistor will operate at while still exhibiting typical characteristics. The typical gain bandwidth may be given in kilohertz (kHz) or in megahertz (MHz).

Finally, most manufacturers will list the type of case or cases in which the device is housed. Certain manufacturers may offer the same device in several different cases. Case style and electronic lead configuration or biasing will usually be described with numbers such as TO-1 to TO-92, etc.

FIELD-EFFECT TRANSISTORS

The *field-effect transistor* (FET) is another type of solid-state device which is often used in electronic circuits. This type of transistor differs from the bipolar designs just discussed in that current flow through the FET is controlled by the variation of an electric field established by a control voltage. The bipolar transistor uses a variation of the current injected into the base terminal to control current flow.

Field-effect transistors exhibit many of the electrical characteristics of common electron tubes. They do this while still retaining the many advantages of solid-state devices, such as mechanical ruggedness, size, and power consumption.

Field-effect transistors fall into two basic groups: *junction-gate field-effect transistors* (JFETs) and *metal-oxide semiconductor field-*

effect transistors (MOSFETs). In both types, an electric field controls current conduction but the electrical characteristics differ.

Figure 2-35 shows schematic representations of two JFETs. These are polarized devices like the bipolar transistor discussed earlier. Each has three leads which are named gate, drain, and source. These electrodes are equivalent to the base, collector, and emitter of bipolar transistors. Figure 2-36 shows schematic drawings for MOSFETs, which are also divided into n-channel and p-channel devices. MOSFETs have four electrodes. The one labeled substrate is an addition to the standard JFET configuration.

MOSFETs are placed into two sub-categories: *enhancement types* and *depletion* types. An enhancement type can be thought of as the equivalent of a normally open switch, while the depletion type is normally conductive or closed. When current is conducted, these devices reverse their normal states. Many transistor checkers will not accurately test the condition of field-effect transistors. Usually, a special type of checker is needed.

DIODES

The *solid-state diode* or *rectifier* is a device which allows the passage of current in only one direction. When current flows through a diode in one direction, it will be opposed in the opposite direction. Due to this characteristic, the diode makes an ideal rectifier. Rectification can be thought of as the changing of alternating current into direct current. The diode accomplishes this by allowing only one cycle of alternating current to pass while blocking the alternate cycle.

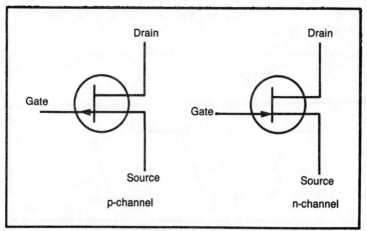

Fig. 2-35. Schematic representation of two junction field-effect transistor types.

49

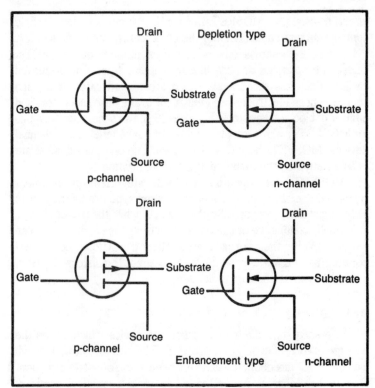

Fig. 2-36. Schematic drawings of the various types of MOSFETs.

When ac current is fed to a rectifier circuit, the output is direct current. Figure 2-37 shows the schematic symbol of a solid-state rectifier. Current flows in the direction of the arrow portion of the schematic symbol.

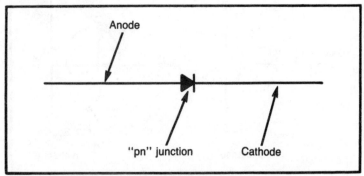

Fig. 2-37. Schematic symbol of a solid-state rectifier.

Diodes are normally constructed of two basic types of semiconductor material: germanium and silicon. *Germanium diodes* are characterized by relatively large current flows with relatively small voltages. They also have a very small current flow in the reverse direction or the direction opposite to that which is indicated by the arrow symbol. Germanium diodes are used mainly for signal-type applications, while *silicon diodes* are used mostly for *rectification* purposes—for converting alternating current into direct current.

Figure 2-38 shows some pictorial representations of various diodes. This is but a very limited representation of the wide assortment of case designs. However, the shape of the characteristic curves of these diodes is very similar. The primary difference is found in the relationships of current and voltage limits. Figure 2-39 shows a typical curve of a junction diode. The graph shows two different kinds of bias. Bias in the pn junction is the difference in potential between the anode (p material) and the cathode (n material). Forward bias is the application of a voltage between n and p material, where the p material is positive with respect to the n material. When the p material becomes negative with respect to the n material, the junction is reverse biased.

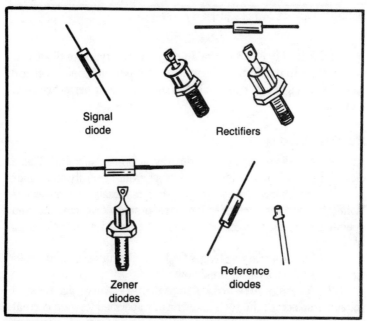

Fig. 2-38. Solid-state rectifiers and diodes come in many different physical packages.

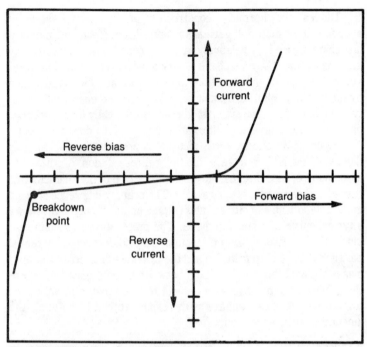

Fig. 2-39. Typical curve of a junction diode.

The following is a description of the major types of diodes. It is important to understand these various types and their electrical characteristics in order to be able to measure their performance in a circuit.

Rectifier Diodes

Rectifier diodes are used primarily in power supplies. These diodes are usually silicon because of its inherent reliability and higher overall performance compared to other materials. Silicon allows higher forward conductance, lower reverse-leakage current, and operation at higher temperatures. The major electrical characteristics are:

1. Dc blocking voltage (V_R). Maximum reverse dc voltage which will not cause breakdown.

2. Average forward voltage drop (V_F). Average forward voltage drop across the rectifier given at a specified forward current and temperature, usually specified for rectified forward current at 60 Hz.

3. Average rectifier forward current (I_F). Average rectified forward current at a specified temperature, usually at 60 Hz with a resistive load. The temperature is normally specified for a range, typically -65 to $+175$ degrees centigrade.

4. Average reverse current (I_R). Average reverse current at a specified temperature, usually 60 Hz.

5. Peak surge current (I_{Surge}). Peak current specified for a given number of cycles or portion of a cycle. For example, ½ cycles at 60 Hz.

Signal Diodes

Signal diodes fall into various categories, including general purpose, high-speed switch, and parametric amplifier types. These devices are used as mixers, detectors, and switches, as well as in many other applications. The major electrical characteristics of signal diodes are:

1. Peak reverse voltage (PRV). Maximum reverse voltage which can be applied before reaching the breakdown point.

2. Reverse current (I_R). Small value of direct current that flows when a semiconductor diode has reverse bias.

3. Maximum forward voltage drop at indicated forward current ($V_F@I_F$). Maximum forward voltage drop across the diode at the indicated forward current.

4. Reverse recovery time (t_{rr}). Time required for reverse current to decrease from a value equal to the forward current to a value equal to I_R when a step voltage is applied.

The schematic diagram for the rectifier and signal diode is shown in Fig. 2-40. Forward current flows into the point of the arrow and reverse current is with the arrow.

Zener Diodes

The *zener diode* is unique because it is designed to operate reverse biased in the avalanche or breakdown region. This device is

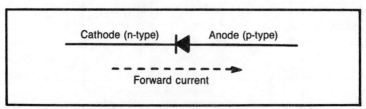

Cathode (n-type)   Anode (p-type)

Forward current

Fig. 2-40. Schematic drawing of a rectifier or a signal diode.

used as a regulator, clipper, coupling device, etc. The major electrical characteristics of zener diodes are:

1. Nominal zener breakdown $V_{Z(Nom)}$. Sometimes a $V_{Z(Max)}$ and $V_{Z(Min)}$ are used to set absolute limits between which breakdown will occur.

2. Maximum power dissipation (P_D). Maximum power the device is capable of handling. Because voltage is a constant, there is a corresponding current maximum (I_{ZM}).

Schematic diagrams of the zener are shown in Fig. 2-41. Zener current flows in the direction of the arrow. In many schematics, a distinction is not made for this diode and a signal diode symbol is used.

Reference Diodes

Reference diodes were developed to replace zener diodes in certain applications because of the zener's temperature instability. Reference diodes provide a constant voltage over a wide temperature range. The important characteristics of this device, besides V_Z, are T_{Min} and T_{Max}, which specify the range over which an indicated temperature coefficient is applicable. The temperature coefficient is expressed as a percent of the change of the reference (V_Z) per degree centigrade in temperature.

Varactor Diodes

Pn junctions exhibit capacitance properties because the depletion area represents a dielectric and the adjacent semiconductor material represents two conductive plates. Increasing the reverse bias decreases this capacitance, while increasing forward bias increases it. When the forward bias is large enough to overcome the barrier potential, high forward conduction destroys the capacitance effect except at very high frequencies. Therefore, the effective capacitance is a function of the external applied voltage. This characteristic is

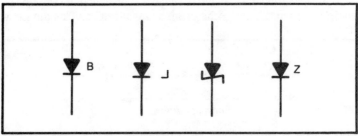

Fig. 2-41. Schematic representation of zener diodes.

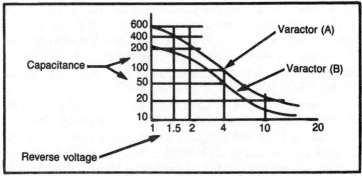

Fig. 2-42. Voltage/capacitance relationship of varactor diodes.

undesirable in conventional diode operation, but is enhanced by special doping in *varactor* or *variable-capacitance (varicap) diodes.* The applications of the varactor can be divided into two main types, *tuning* and *harmonic generation.* Different characteristics are required by the two types, but both use the voltage dependent junction capacitance effect. Figure 2-42 shows the voltage of capacitance relationships. The use of this diode for frequency multiplication is common.

As a variable capacitor, the varactor is rugged and small. It is not affected by dust or moisture, and is ideal for remote control and precision fine tuning. The current uses of tuning diodes span the spectrum from AM radio to the microwave region. The most significant parameters of a tuning diode are the capacitance ratio, Q series resistance, nominal capacitance, leakage current, and breakdown voltage.

The *capacitance ratio,* which defines the tuning range, is the amount of capacitance variation over the bias voltage range. It is normally expressed as the ratio of the low-voltage capacitance divided by the high-voltage capacitance. For example, a typical specification which reads $C_4/C_{60} = 30$ indicates that the capacitance value at four volts is three times the capacitance value at 60 volts. The high voltage in the ratio is usually the minimum breakdown voltage specification. A four-volt lower limit is quite common, since it describes the approximate lower limit of linear operation for most devices. The capacitance ratio of tuning diodes varies in accordance with construction.

SILICON CONTROLLED RECTIFIERS

The *silicon controlled rectifier* (SCR) is a solid-state device which is made up of layers of crystalline material which have been treated so as to turn the material into a semiconductor. The SCR is a special kind of diode which allows current to pass through the device upon a command from an external signal. This signal is applied between the anode and the third element, called the *gate*. Aside from the capability to switch in this manner, the SCR is similar to a standard diode. The SCR's output is direct current.

Shown in Fig. 2-43 is an example of a circuit which uses the SCR both as a standard rectifier of ac and as a switch. The input is house current which is rectified by the switched-on SCR into the ac equivalent voltage in direct current. This is a half-wave rectifier circuit because only one-half of the ac sine wave is acted upon by the single SCR. The switching circuit is composed of a small resistor connected between the gate and the anode. When this resistor is switched into the circuit, a small amount of current flows between the gate and anode. The result of this action is that the SCR begins conducting current. Until this time, the SCR appears as an open circuit and no current is delivered to the load.

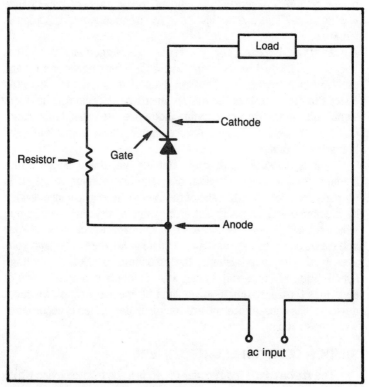

Fig. 2-43. A control circuit using an SCR.

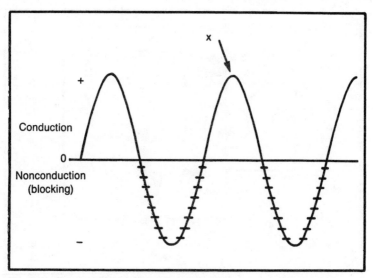

Fig. 2-44. Graph showing conduction and nonconduction curves as they apply to rectifier operation.

The SCR will continue to conduct current as long as the switch is left in the on position. When the switch is turned off, the current flow will appear to stop immediately. In this circuit, for all practical purposes, it does. However, the SCR will not cease conducting until the input current reaches zero. This occurs every 1/120th of a second in a 60-cycle ac circuit, which is used as the source of electrical power in this circuit. Thus, if the switch were thrown off during the peak of one cycle in which the rectifier was conducting, the device would not stop conducting until the cycle completed its decay to zero.

This can be understood further by referring to Fig. 2-44. Here, the ac sine wave can be seen as it would appear on an oscilloscope. This is a half-wave rectifier, which, as already mentioned, only conducts during one-half of the cycle. In this case, it is conducting during the portion of the cycle indicated by the curves on the top portion of the zero line, or the positive cycle. During the other half of the cycle, it is blocking the flow of current. If the switch were thrown during the part of the cycle marked x, the SCR would not stop conducting until the cycle peaked and returned to zero.

Although this 1/120th of a second lag may seem insignificant, it becomes quite important when a dc power source is being used, along with the SCR for control and switching. Figure 2-45 shows the same basic circuit with dc substituted for ac. Since a rectifier will pass the flow of current in only one direction, it will easily conduct

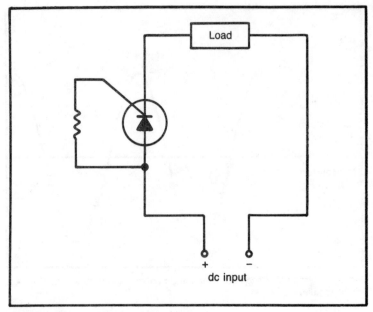

Fig. 2-45. SCR circuit using dc voltage.

dc which is of a fixed polarity. This assumes that the diode is connected with the correct polarity as in this circuit. To reverse the diode would result in a zero output current.

As before, a small resistor and switch set up a circuit between the gate and the anode. When the switch is thrown, the SCR will fire and current will be conducted through the load. When the current is switched off, there will be no change, because the SCR will continue to conduct until the current it is passing drops to near zero. The ac circuit did this several times in a fraction of a second, but dc stays at a constant value at all times. The result is that in a circuit with a dc power supply, the SCR will continue to conduct, regardless of the position of the gate/anode switch, until the source of current is removed from the circuit. At this time, the SCR will quickly return to its nonconducting state and it will be necessary to turn the gate/anode switch to the on position to return it to a conducting state again.

TRIACS

A *triac* is another semiconductor device which operates in a somewhat different manner than the SCR already discussed. A triac, when triggered, will pass current in either direction. It is ideal for the control of ac circuits. You already know that an SCR's output will always be direct current. This is suitable if the load requires dc. However, if it is desirable to supply alternating current to a load, an SCR will not do. Figure 2-46 shows the use of a triac to provide just this. Here, two SCRs are combined in a reverse parallel arrangement. The anode of one is connected to the cathode of the other and the cathode of the first is connected to the anode of the second. The gates are tied together. In this way, current of one polarity will be conducted through one SCR; when the current reverses, it will be conducted through the other. Both diodes are located within the same major circuit, so alternating current will be passed by this combination device when both are triggered. Fortunately, it is not necessary to use two SCRs to perform in this manner. The triac is a single, compact component which is about the size of a single SCR. The circuitry is connected internally through a chemical semiconductor manufacturing process.

Figure 2-47 shows a simple circuit that uses the triac to control current flow through an alternating-current load. A small-wattage, high-value resistor is used as the triggering mechanism in combination with a small single-pole, single-throw switch. The triac appears as

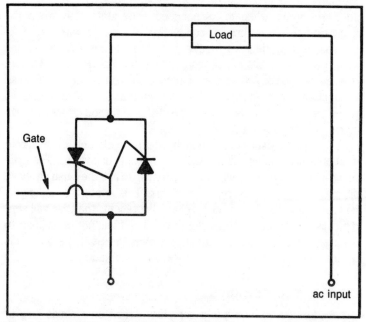

Fig. 2-46. Two SCRs combined in reverse-parallel form a device known as a triac.

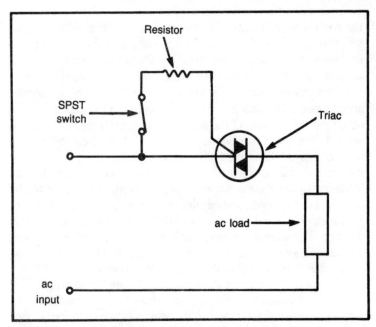

Fig. 2-47. Triac circuit used to control current flow through an alternating-current load.

an open circuit when in its nonconducting state; but when the triggering switch is thrown, conduction begins. As with the SCR, the triac will continue to conduct until the current source reaches zero. Remember that in ac circuits, this action occurs every 1/120th of a second, so the switch may be turned off and the circuit will cease conduction within an instant. The switch must be left closed at all times for this reason, because if the switch is turned off, conduction ceases almost immediately.

The triac is usually an expensive device, although it is available in a configuration that will handle large amounts of current. The main thing to watch out for when using triacs is that they usually have specific voltage and current limitations. This is not a real danger, however, because if too much voltage is present, the triac will simply cease to conduct. Thus, the device itself will not be subject to damage. This is a distinct advantage over the SCR, because this device may be destroyed under the same conditions.

INTEGRATED CIRCUITS

An *integrated circuit* is a complete circuit on a single chip of semiconductor material, usually silicon. It is composed of the same types of materials used to manufacture the discrete devices already discussed, such as transistors, diodes, etc. An integrated circuit may consist of these same devices all housed in one compact unit. The amazing thing about an integrated circuit is that although it may contain many transistors or other solid-state devices, resistors, and other components, it will usually only weigh a fraction of an ounce and can be held between the thumb and forefinger of your hand. An integrated circuit may contain over 1000 solid-state devices, and this miniaturization has revolutionized the electronics industry.

In an integrated circuit, all of the components used must be processed from the same solid-state material. The manufacturing process may differ somewhat from company to company, but the basic process remains the same. The silicon or other material is treated with impurities to create a semiconductor. The various types of semiconductor materials are combined to form the different components. The chip of silicon used in an integrated circuit is simultaneously treated in a manner which causes microscopic portions of the water to be transformed into the different components. One tiny part, then, acts as a transistor, for example, while another will be a diode or a resistor.

An integrated circuit can be wired to produce several different circuits, depending on the wiring configurations used. Shown in Fig. 2-48 is an integrated circuit as it would appear if discrete components were used instead. A circuit made with discrete components would be much larger than an integrated circuit. This enormous savings in physical space is responsible for the compact size of many types

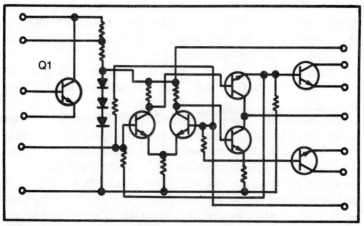

Fig. 2-48. An integrated circuit as it would appear if built from discrete components.

of electronic equipment available today. It is also important to note that one integrated circuit can be treated as a discrete component itself when combined with other parts of a larger circuit. The only difference is that instead of thinking in terms of an individual component, you must think in terms of completed circuits on tiny chips.

The integrated circuit is known for its dependability and ruggedness. By using integrated circuits without any external, discrete devices, dependability can be increased even further. The greatest reliability is obtained when a single integrated circuit is used with a minimum of external connections or devices. ICs are impact-resistant, not subject to vibration damage, and can be enclosed in a compact case for protection.

The difference in the cost of an integrated circuit when compared with the cost of purchasing the same discrete devices to make up the same circuit is significant. An integrated circuit may cost as little as $2.00, while the discrete devices, if purchased individually, could be as much as $500. This makes the integrated circuit much more desirable for not only large manufacturers of electronic equipment, but for the home experimenter as well. It is easy to see why the integrated circuit has quickly replaced many discrete components in instances where it is practical.

LIGHT-SENSITIVE SOLID-STATE DEVICES

Although solid-state devices have been available for many years, their light-sensitive counterparts will not be as familiar to some readers. The light-sensitive devices, as their name implies, respond directly to the presence of light. In other words, instead of responding to a source of current, these devices are triggered by a source of light, either natural or artificial.

Light-sensitive, solid-state devices are finding applications in many electronic circuits today. They can serve to measure the intensity of light, as daylight warning switches, or even for communications purposes. Light-sensitive components fall into two major categories, *photo-conductive* and *photo-emissive*, depending upon the effect which occurs when they are exposed to a light source. An example of a photo-emissive device is the solar cell, which generates electrical current when light strikes its surface. In a photo-conductive device, the resistance to electron flow is changed by the amount of light which strikes its surface. The cadmium sulfide cell is a good example of a photo-conductive device. Those which generate electrical current in response to light can be thought of as active, while those which change their resistance to the flow of electrons are passive. There are many types of light-sensitive, solid-state components available today for a variety of applications.

Solar Cells

As stated previously, the *solar cell* falls into the photo-emissive category of light-sensitive solid-state devices. The solar cell is sometimes referred to as the *photoelectric cell* and can be directly utilized to power a circuit which requires low voltage and current. Although the shape may vary, a solar cell usually consists of a flat plate with a specially treated surface. This device is sometimes likened to a dry cell battery because it consists of a positive and negative terminal and can be connected in a parallel circuit for increased current, of in a series circuit for higher voltage output.

Figure 2-49 shows the schematic representation of a solar cell. These devices normally have an output of approximately 0.45 volt. The output from the solar cell is dependent upon the amount of light which is present at its sensitive surface. Although the cell has a maximum output, greater light levels or intensities will create higher current ratings until this maximum is reached. The light energy is transformed into electrical current.

As might be expected, the solar cell has quite a high price tag, although in recent years the price has dropped considerably. This reduction can be attributed to technological advances with regard to manufacture and efficiency. It is certain that with further research, the price will become much more competitive, making the solar cell a viable alternative to their more standard counterparts.

Cadmium Sulfide Photocells

You already know that a resistor is a device which hinders the flow of current, and a conductor is one which passes current. Any

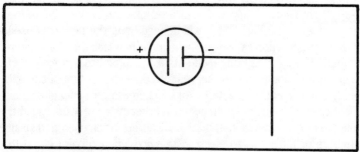

Fig. 2-49. Schematic representation of a photovoltaic or solar cell.

material that is used to fabricate electronic components will offer both resistance and conductance to the flow of current. The *cadmium sulfide cell* is a special semiconductor device which offers a resistance to the flow of current. This cell is similar in operation to a volume control, except that instead of a manual resistance control, the cadmium sulfide cell responds to the amount of light which strikes its surface. In other words, in the absence of light, the cadmium sulfide cell will exhibit the greatest amount of resistance; while when it is struck by light of high intensity, resistance will become lower and lower until the device reaches a specified minimum resistance. Because of this action, the cadmium sulfide cell is of the photoconductive type.

Because the cadmium sulfide cell does not produce any electrical current on its own, it is normally used in circuits that utilize another source of electrical power. It may also be used in conjunction with a solar cell in a circuit. The result would be a circuit which is not only powered by light but controlled by it as well. The cadmium sulfide cell is used for control through placement in the base leg circuit of a bipolar transistor to turn it on and off. When light strikes the surface of the cell, the resistance changes, thus varying the amount of current which is delivered to the base lead. The transistor will then conduct or change to another state. This circuit can be used to trigger a relay or activate other types of electronic circuits. Fig. 2-50 shows a circuit with its own power supply. When light strikes a cadmium sulfide cell, the transistor conducts and completes the circuit through the relay, which is used to active other pieces of equipment.

Light-Emitting Diodes

The *light-emitting diode* is composed of semiconductor material which has been treated with gallium arsenide. This device has the property of emitting light in the infrared spectrum in addition to the visible spectrum. This ability will depend upon the materials used in their construction as well as the manner in which they have been treated. The light-emitting diode or LED can be had in a variety of sizes, shapes and forms and serves to reduce the amount of current needed in an electronic device. An LED emanates a cool glow that is not created by heating effects as in other types of artificial light; hence, energy consumption is less. LEDs are often used to display numbers or letters in an electronic device when current is fed to its contacts.

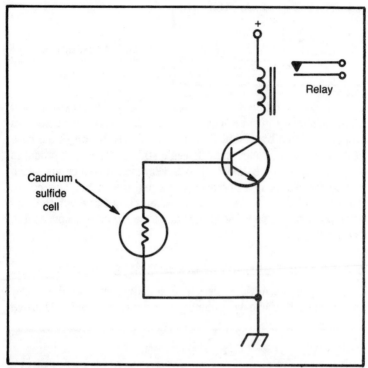

Fig. 2-50. A cadmium sulfide cell may be used to control switching of a transistor and relay circuit.

Although an LED is not technically a light-sensitive device, it is often used in combination with one of those already discussed, such as the solar cell or the cadmium sulfide cell. In this type of arrangement, either one or more LEDs are positioned so that their light rays strike the surface of the light-sensitive device, thus causing a conversion to take place several times. Direct current is passed through the LED and converted to light. This strikes the surface of the light-sensitive device, thus controlling the flow of current in another circuit in the case of the cadmium sulfide cell and generating current if the solar cell is used. Figure 2-51 shows a light-emitting diode.

Phototransistors

Photocells are composed of a junction made when two types of semiconductor materials are pressed together in such a way that when light strikes near the junction, a flow of electrons is released. If this transparent junction is backed up by another semiconductor

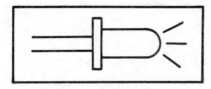

Fig. 2-51. A light-emitting diode.

layer called a collector, the result is the *phototransistor*, and *photoconductive current* generated by the photocell junction is amplified as in a transistor. The sensitized area becomes the base junction of the transistor and when light strikes this area, it produces an input signal to the device. This is very similar to a circuit which can be formed using a photocell and a pnp transistor. This circuit is shown in Fig. 2-52. The phototransistor does away with the separate photocell or photodiode portion of the circuit and houses it in one unit.

OTHER LIGHT-SENSITIVE SOLID-STATE DEVICES

Some of the devices discussed already , such as the diode, transistor, and triac have a number of derivatives which will prove

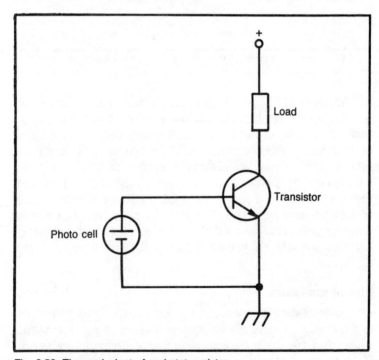

Fig. 2-52. The equivalent of a phototransistor.

66

useful in specialized situations. In applications where it is desirable for a piece of electronic equipment to be light-activated, these three devices can be obtained in light-controlled versions. Aside from the fact that the diode, transistor, or triac is activated by a source of light, they will operate in much the same manner as their conventional counterparts.

A *light-activated, silicon-controlled rectifier* (LASCR) is shown in Fig. 2-53. This device will consist of the three main elements described earlier in regard to the standard SCR, but note in this schematic that there are two arrows pointing to the schematic symbol for this device. This indicates light activation, and the LASCR will not conduct when no light strikes its surface. It will only trigger when it is exposed to a source of light. As before, once the device is triggered, it will cease to conduct only when the input current gets close to zero. Operation of the LASCR, then, is basically the same as the conventional SCR.

Shown schematically in Fig. 2-54 is the light-activated transistor or phototransistor. Notice that this closely resembles that of the bipolar transistor, except the base lead is often removed. The base or biasing input to the phototransistor is in the light rays which strike the light collector. Here, the intensity of the light causes the device to conduct electricity. In other applications, the transistor can form the input to an audio oscillator which will vary in tone as the light level increases or decreases. With all three light-activated devices,

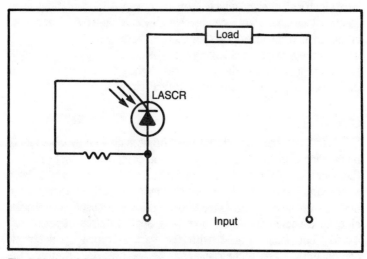

Fig. 2-53. A LASCR is used just like the silicon-controlled rectifier, except light is used as the controlling element.

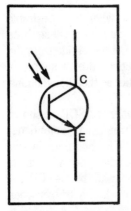

Fig. 2-54. A light-activated transistor shown in schematic form.

light rays can thus be used as the prime controlling element in an electronic piece of equipment.

TRANSDUCERS

A *transducer* is a device which takes one form of energy and converts it to another form. A good example of a transducer is a microphone. A microphone receives audible energy and converts it into current flow in an electrical circuit. Another example is a stereo speaker. Here, the current flow in a circuit is converted back into audible sound so it can be heard by the human ear. Photo-emissive devices are also transducers. They convert light energy into electrical energy which then produces a current flow in an electrical circuit. Photo-conductive devices, however cannot be classified as transducers. Rather than converting energy from a light source, they react by changing their internal resistance. It should be apparent from this discussion that light can be used as a form of energy in the operation of electrical circuits.

SUMMARY

When performing any test function, it is absolutely essential to be familiar with all electronic values and components that relate to the operation of the equipment under test. You do not need to have a thorough background in electronic theory to perform these tests, but you do have to be familiar with the terms used to describe electronic values and know how they apply to circuit operation.

By becoming familiar with the various types of electronic components, the technician can better decide whether the measurements indicate a possible problem area. In an electronic circuit, components interact with each other and can cause unusual readings to occur at a test point which is nowhere near the problem area. By understanding this interaction, the technician is in a much better position to make use of the readings obtained to pinpoint potential problem areas by a process of elimination and deduction.

Understanding electronic values and components is not a difficult task. Only a few hours of study are actually required to gain a basic grasp of their operation and meaning. Then, through experience in actual bench testing, the basic knowledge you have gained will be magnified by practical application. Some beginning technicians feel that experienced troubleshooters can identify a particular equipment problem simply by observing the operating symptoms of a particular piece of equipment. This is not true in many cases. Correct troubleshooting involves coupling a basic understanding of electronic values and components with a series of measurements at proper circuit points. Then, by comparing these obtained readings with a knowledge of how components should be operating, an elimination process is begun which will sooner or later lead to the exact problem. Speed in doing this is obtained only through continual study and actual practice at the electronics test bench.

Chapter 3

Circuit Testing Considerations—
The Logical Approach

The information provided in the two previous chapters laid the foundation for measuring circuits and equipment by describing units of measurement, the principal components found in electrical and electronic devices, and their values. This chapter will discuss a logical and sequential method of both troubleshooting and maintaining electronic equipment which will make the process simple and efficient.

The term, *troubleshooting* is one which will be heard and used very frequently in the field of electronics. However, troubleshooting is sometimes misinterpreted to mean simply fixing a piece of equipment when it fails. This is only part of the overall picture. In addition to this step, the troubleshooter must be able to evaluate equipment performance by comparing theoretical knowledge with the present indications of performance characteristics.

The term, *maintenance* refers to all actions which a person performs on an actual piece of equipment or machinery to retain the equipment in a serviceable condition or to restore it to serviceability. This involves inspection, testing, servicing, repair, rebuilding, etc.

A logical or systematic approach to troubleshooting is paramount. Needless to say, many hours of time have been lost due to the "hit or miss" type of approach. It is necessary to develop a concept for troubleshooting, which, if followed, can give you the ability to repair any piece of electronic equipment, regardless of its complexity or purpose. This chapter will discuss six basic steps and operations in troubleshooting that will save you many hours of wasted time over the trial and error approach. These six steps are:

1. Symptom recognition.
2. Symptom elaboration.
3. Listing of probably faulty functions.
4. Localizing the faulty function.
5. Localizing trouble to the circuit.
6. Failure analysis.

SYMPTOM RECOGNITION

The first step in a logical approach to troubleshooting is to first determine whether the equipment is functioning properly. Any piece of electronic equipment is designed to perform a specific job or group of jobs as specified by certain performance parameters. What this means is that a certain type of performance should be obtainable at all times. If it were impossible to know when the equipment was performing poorly, it would be equally impossible to maintain the equipment in satisfactory condition. The first step, then, is to recognize a malfunction in the equipment.

Since a symptom is a manifestation of an undesirable change in equipment performance, it is necessary to have some standard of normal performance as a reference. By comparing the present performance with this standard, it is possible to recognize a symptom and to make a decision as to just what the symptom is. For example, the normal television picture is a clear, properly contrasted representation of an actual scene. It should be centered within the vertical and horizontal boundaries of the screen. If the picture suddenly begins to "roll" vertically, this is easily recognized as a symptom because it does not correspond with the normal operation which is expected. In the same way, the normal sound from a receiver is a clearly understandable reproduction of the message sender's voice. If it sounds as though the sender is talking from the bottom of a barrel filled with water, this is an indication that distortion is occurring. This, again, is a symptom.

Of course, if an electronic piece of equipment fails to operate at all, this is easily recognizable. Failure of operation means that either the entire device or else some part of the equipment is not functioning. For example, the absence of sound from a receiver when all controls are in their proper positions indicates a complete or partial failure.

Even if both the auditory and visual signals are present, however, a piece of equipment may not be performing as it should. If the equipment is working but is presenting information that does not meet design specifications, the performance is said to be degraded. This situation should be corrected just as quickly as an equipment failure. Performance may range from a nearly perfect to barely operating condition.

It is important to remember that any electronic equipment, no matter how complex, is built by using the basic electronic components discussed in the first two chapters in this text. These components are combined in such a way that the desired performance is produced.

SYMPTOM ELABORATION

The second logical step in troubleshooting is to further define the obvious or not so obvious symptoms. Most larger electronic devices or systems have operational controls, additional indicating instruments (other than the main one), or other built-in aids for evaluating performance. These should be utilized at this point to see whether they will affect the symptom which has been recognized or provide additional data that further defines the symptom.

It is not recommended that test equipment be used at this point in the troubleshooting procedure. Unless you completely define the symptoms first, you can quickly and easily be led astray. The result would be loss of time, unnecessary expenditure of energy, and perhaps even a dead-end approach. This step, then, is the "I need more information" portion of a logical approach to troubleshooting.

To illustrate the necessity for this step, consider the following situation. Recognizing an undesirable hum in a receiver could lead you in several directions if you do not obtain a more detailed description of the symptom. Receiver hum can be due to a number of things, such as poor filtering action in the power supply, heater-cathode leakage, ac line voltage interference, or other internal and/or external faults. It should be apparent, then, that the primary reason for elaborating further is that similar symptoms can be caused by more than one malfunction. In order to proceed efficiently, it is necessary to make a valid decision as to which fault or faults are probably producing the specific symptom in question.

On equipment which is equipped with operating controls, these can be used to further define the problem. The operating controls are considered to be all front-panel switches, variable circuit elements, or mechanical linkages connected to internal circuit components that can be adjusted without going inside the equipment enclosure. They control the supply of power to the equipment's circuits, tune or adjust various performance characteristics, or select a particular type of performance.

By their very nature, operating controls must produce some sort of change in circuit conditions. This change will indirectly alter current or voltage values by the direct variation of resistance, inductance, and/or capacitance elements in the equipment circuitry. The information displays associated with the equipment (front-panel meters and other indicating devices) will enable you to "see" the changes that take place when the controls are operated.

Control manipulation can cause detrimental effects in equipment performance as well as the desirable effects for which they are primarily intended. Manipulation of the controls in the wrong order or having allowed voltage and current values to exceed maximum design specifications are often the cause of the original problem. Unless you observe the proper precautions while investigating the symptom, even more damage to the equipment can result from the improper use of operating controls.

Every electronic circuit component has definite maximum current and voltage limits below which it must be operated in order to prevent it from insulation breakdown and other types of damage. The meters placed on the front panel serve as an aid in determining voltage and current values at crucial points in the equipment circuitry. Operating controls should never be adjusted so that these meters indicate values above the maximum ratings.

In addition to exceeding maximum ratings and manipulating controls in an improper sequence, there are certain other precautions associated with specific types of equipment. For example, the intensity control on an oscilloscope (which will be discussed in detail in another chapter) should never be adjusted to produce an excessively bright spot on the fluorescent screen. A bright spot indicates a high current which may burn the screen coating and decrease the life of the tube. Also (and for the same reason), you should never permit a sharply focused spot to remain stationary for any length of time.

Another precaution concerns the adjustment of the range selector switch on any type of indicating meter. If the switch is carelessly positioned to a range below the value of the quantity being measured, the needle will strike its upper mechanical limit. This may bend the needle and result in inaccurate (offset) readings. Of course, it would be impossible to include in this text each and every precaution associated with equipment and measuring devices. The instruction and performance manuals should always be consulted for details.

It is important to emphasize that incorrect control settings may produce an *apparent* problem. The word "apparent" is used because the equipment may be operating perfectly, but because of the incorrect setting, the information obtained will not correspond with the expected performance. An incorrect setting may be brought about by an accidental movement of the control as well as by careless misadjustment.

Let's assume you are using an oscilloscope to check the voltage across the load resistor of an audio amplifier stage in a superheterodyne receiver. The waveform should be 100 volts from the positive peak to the negative peak, and you are trying to verify this amplitude in order to evaluate the performance of the amplifier stage. You intend to set the vertical sensitivity switch to 50 volts per centimeter and expect to see a display waveform similar to that in Fig. 3-1. However, in haste, you accidentally set the vertical sensitivity switch to 10 volts per centimeter. As a result, the display is instead similar to that in Fig. 3-2. Thus, at first glance, you will think that the amplitude is 500 volts because you assume that the switch is set at 50 volts per centimeter. Certainly, your first thought will be that the amplifier is not functioning properly.

At this point, a knowledge of amplifier operation should be applied. Figure 3-3 is the circuit diagram for the amplifier under test. It should be apparent that because the supply voltage for the amplifier

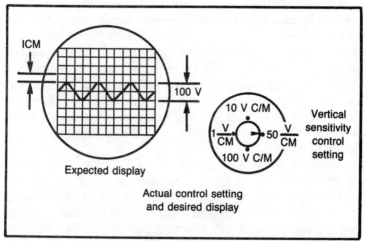

Fig. 3-1. With the vertical-sensitivity switch set as shown, the display would look like this.

74

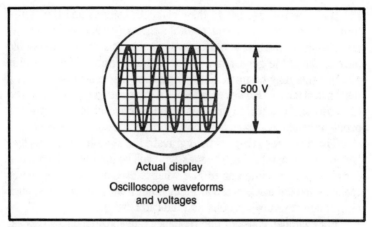

500 V

Actual display

Oscilloscope waveforms
and voltages

Fig. 3-2. With the vertical sensitivity switch set in error, the display will appear similar to this.

is only 150 volts, it is an operational impossibility for 500 volts to exist across the load. The amplifier shown in this figure cannot produce an output voltage larger than its own plate supply voltage.

The next logical assumption is that the oscilloscope is in error. Because it is the vertical dimension that is apparently causing the problem, you should direct your attention to the vertical sensitivity control. This will result in your detecting the error in the control setting.

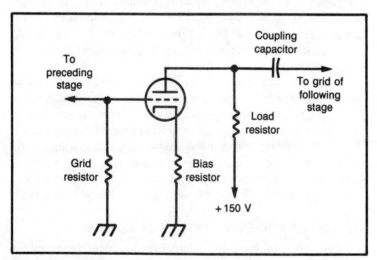

Coupling
capacitor

To
preceding
stage

To grid of
following
stage

Load
resistor

Grid
resistor

Bias
resistor

+ 150 V

Fig. 3-3. Circuit diagram for the audio-amplifier stage in a superheterodyne receiver.

If all controls are set to their correct positions and the symptom still persists, it is still possible that an operating control is responsible for the problem. However, in this case, the trouble would have to fall in the general area of component failure. If a control is at fault, this may be immediately apparent, especially when it is a mechanical failure. However, additional information may be required to determine when a control has failed electronically, because the problem produced may also point to other electronic failures.

The next step, then, is to aggravate the symptoms if possible. By observing any further changes, you will be able to make a more accurate determination as to just what is causing the panel meter readings and the displays produced by other front-panel devices, such as cathode-ray tube screens and indicator lights.

For example, consider the frequency range switch on a broadcast transmitter. This control is a multiposition switch with each position connecting a different rf coil in parallel with the main tuning capacitor of the oscillator tank circuit. The value of each coil is such that it will cause the oscillator to vary over a different range of frequencies as the tuning capacitor is varied. If a weak transmission symptom is detected, it is logical to try other frequency ranges by manipulating the selector switch. If normal transmission is achieved for any of the range conditions, the fault probably lies either in the switch itself or in a few of the tuning coils. Such an observation would provide a quick location of the problem area.

Similarly, in a receiver, if a mode selector switch can be changed from AM operation to FM operation, it is logical to check the receiver in both positions. If the symptom persists only in the AM mode, the circuitry associated with the FM mode can be eliminated from consideration.

The next step is to gain further information about a trouble symptom by manipulating the operating controls and other instruments. If the trouble is cleared up by manipulating the controls, the analysis may stop at this point. By applying a knowledge of the equipment, though, you should still find the reason why the specific control adjustment removed the apparent malfunction. This action is necessary to assure yourself that there are no additional faulty components which will produce the same problem at a later time.

LISTING OF PROBABLE FAULTY FUNCTIONS

This step does not apply to smaller and less sophisticated pieces of electronic equipment which contain only one functional area or unit. On larger pieces of equipment, there may be many, many circuits and individual components. Because of this, it is necessary to break the equipment down into separate units to isolate the area of the malfunction. The term *functional area* is used here to denote an electronic operation performed by a specific area of the equipment. A transceiver, for example, may include a transmitter, modulator, receiver, and power supply. The combined functions of these units allow the equipment to perform the electronic purpose for which it was designed. A functional unit may be located at one or more physical locations and consists of all the components that are required for the unit to perform properly.

Once the equipment has been broken down into functional units, it is necessary to logically determine which one could be responsible for the problem. In this way, it will be possible to narrow the field down and make a list of all the probable causes.

In a relatively simple piece of equipment with only one functional unit, the process will be a bit less complicated. Figure 3-4 shows the circuit for a multirange ohmmeter used to make resistance checks. It consists of a current indicating meter, a battery, and resistors of known value connected so that the unknown resistance can be compared with one of the known resistors. When the test

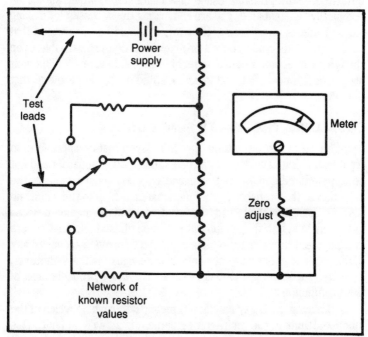

Fig. 3-4. Circuit for a multirange ohmmeter used to check resistance.

leads are open, no current flows through the meter, and the meter is mechanically set to indicate an infinite resistance. When the test leads are shorted together, the meter is electrically adjusted to give a zero reading.

When the test leads are placed across an unknown resistance, this resistance is in series with the battery and meter. The meter is both in series and in parallel with the known resistors. Thus, the current through the meter will be some intermediate value between those values which produced zero reading and those that produced an infinite reading. The actual value of this current will be determined by the ratio of the fixed and unknown resistors. An appropriately calibrated scale will allow the meter to indicate the value of the resistor being measured. As mentioned earlier, with equipment of this simple type, it will not be necessary to go through the process of separating the system or systems into separate units and making a list of the probable units which could cause the malfunction. The next step, that of localizing the faulty function, may also be omitted.

With larger pieces of equipment, both this step and the next are essential. One of the reasons for performing these steps is to save time, although it may not be obvious to you at this point. By determining the probable cause of a malfunction, which should be technically substantiated, you are eliminating the necessity for making illogical checks on all units. However, it should be understood that each unit so selected is only a probable source of the trouble even though its selection is based on technically valid evidence. The next step uses a time-saving and logical method of locating the unit that is faulty.

LOCALIZING THE FAULTY FUNCTION

The first three steps in this systematic approach to troubleshooting have dealt with the examination of apparent and not so apparent equipment performance deficiencies, as well as the selection of the probably faulty functional units. Up to this point, no test equipment other than the controls and indicating devices physically built into the equipment has been utilized. No dust covers or equipment drawers have been removed to provide access to any of the parts or internal adjustments. After evaluating the symptoms, you have made certain decisions about the most probable area of the malfunction.

Localizing the faulty function(s) means determining which of the functional units of the multi-unit equipment is actually at fault. This is accomplished by systematically checking each possible faulty unit until the defective one is found. If none of the units in your list of possible problems is working improperly, it will be necessary to backtrack and reevaluate the symptoms.

At this point, you will bring into play your factual equipment knowledge and skill in testing procedures. The utilization of standard or specialized test instruments and the interpretation of the test data will be very important throughout this and the remaining troubleshooting steps.

A number of factors should be considered before applying any test equipment to an individual unit. Again, knowledge of the equipment as well as access to its performance specifications, should be used to aid you in determining which functional unit should be tested first.

Test-point accessibility is also an important factor to be considered in this decision. A *test point* is a special jack located at some accessible point on the equipment, such as the front panel or chassis. The jack is electrically connected (directly or by means of a switch) to some important operating potential or signal voltage. Actually, any point where wires join or where components are connected can serve as a test point.

Past experience and a history of repeated failures will be quite helpful at this point. This should have some bearing on which probable cause of the problem should be tested first. However, undue attention should not be paid to these factors unless they relate to the logical conclusions arrived at in the previous steps.

By applying the appropriate test instrument, you should obtain either satisfactory or unsatisfactory indications. If the final check of all probable faulty units does not pinpoint the problem, this indicates that you have either made an error in performing one of the tests, or the results of the test were misunderstood. If this occurs, it is necessary to retrace all steps and repeat any which may be doubtful for either of these reasons. Once rechecking everything, if all suspected units still prove to be operating satisfactorily, you should return to the second step and further define the problem.

Having isolated the fault to the actual functional unit, it is now necessary to consider whether a fault in this unit could logically produce the trouble symptom and fit the other information obtained during symptom elaboration. For example, in Fig. 3-5, an AM transceiver has been divided into functional units. Assume that the fault lies in the antenna assembly unit, which does not automatically switch over to the receiver function as it should. First of all, this symptom should occur only in the units associated with the receiver function.

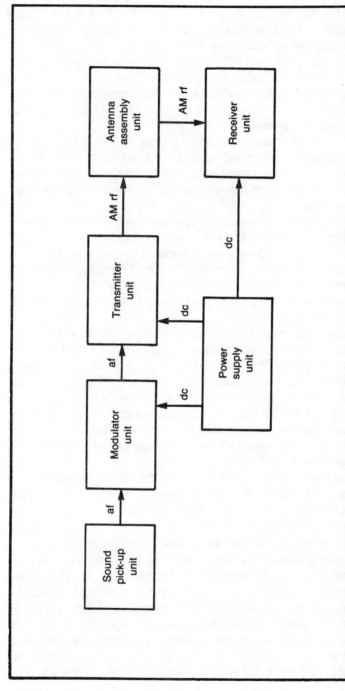

Fig. 3-5. An AM transceiver which has been divided into functional units for ease in troubleshooting.

This would include the receiver unit. The modulator and transmitter should be performing properly. The receiver should provide all normal responses, noise in the speaker, and the ability to vary the noise with the operating controls. However, no signal would be present. If the original symptom and the data collected during symptom elaboration fit the above expectations, you have verified the faulty unit.

The completion of this step should leave no doubt as to which functional unit is at fault. However, as a final check of your work to this point, I advise that you backtrack once again and match the theoretical symptoms to those actually present.

LOCALIZING TROUBLE TO THE CIRCUIT

In this step, it will be necessary to perform some testing with the appropriate test instruments in order to isolate the trouble to a specific circuit within the faulty functional unit. Of course, in the case of a piece of equipment with only one unit, it is still the same basic procedures. Depending upon the complexity of the device, this step may take a very short time or may take much longer. If the equipment is quite large and complex, probably the easiest approach would be to further break down the functional unit into subdivisions and then narrow down the probable causes even further. In order to do this, you must be able to recognize a circuit group. A circuit group is one or more circuits which form a single functional division of a functional unit of a piece of equipment. For example, a typical radio receiver contains the following circuit groups: rf amplifier, converter, i-f amplifiers, detector, and audio amplifiers. A typical radio transmitter contains a master oscillator, intermediate power amplifiers, and final power amplifier. It can be seen from these examples that the circuit groups perform subfunctions and that their combined operations perform the complete function of the unit they make up.

The signals associated with a circuit group normally flow in one or more of four different types of *signal paths*. These include the linear path, convergent-divergent path, feedback path, and switching path. The *linear path* is a series of circuits arranged so that the output of one circuit feeds the input of the following circuit. Thus, the signal proceeds straight through the circuit group without any return or branch paths. This is shown in Fig. 3-6.

The *convergent-divergent path* may be any of three types: divergent, convergent, and the combined convergent-divergent. A *divergent path* is one in which two or more signal paths leave a circuit,

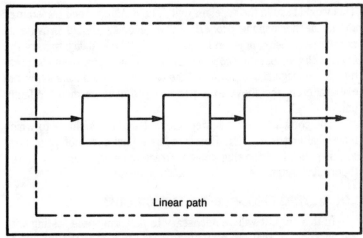

Fig. 3-6. In a linear path, the output of one circuit feeds the input of the next circuit.

as shown in Fig. 3-7. When two or more signal paths enter a circuit, this is known as a *convergent path*. An example is shown in Fig. 3-8. A *convergent-divergent path* is one in which a circuit group or single circuit has multi-inputs and multi-outputs, as shown in Fig. 3-9. This type is not as common as the convergent and divergent paths. The feedback path shown in Fig. 3-10 is a signal path from one circuit to a preceding circuit. The switching path in Fig. 3-11 has a switch for different signal paths.

The subfunctions performed by the various circuit groups and the signal paths interconnecting these groups should be considered in the selection of possible faulty circuit groups. The inputs and/or outputs of these groups should be checked, with the most probable subfunctions being checked first. The accessibility of test points is an important consideration when checking these circuits. As each test is performed, the results should be noted so that they may be referred to if necessary. This procedure should serve to isolate the faulty circuit. The next step, failure analysis, concerns locating the defective part in the faulty circuit.

FAILURE ANALYSIS

This final step will require that certain branches of the faulty circuit be tested in order to determine where the faulty part lies. These branches are the interconnected networks associated with each element of the transistor, electron tube, or other active de-

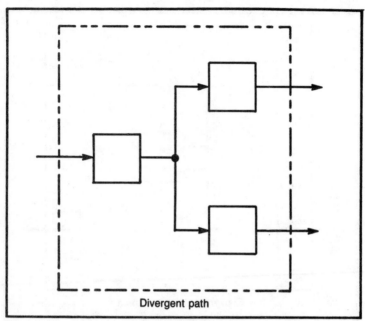

Divergent path

Fig. 3-7. When two or more signal paths leave a circuit, it is called a divergent path.

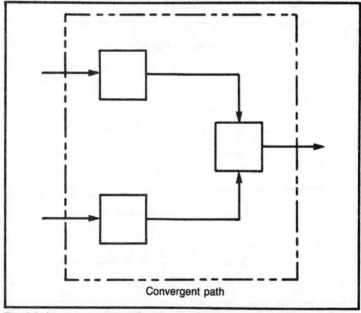

Convergent path

Fig. 3-8. In a convergent path, two or more signal paths enter a circuit.

83

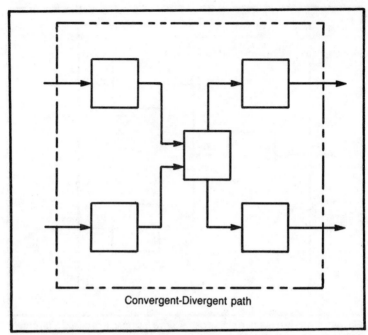

Convergent-Divergent path

Fig. 3-9. Multiple inputs and outputs in a circuit classify it as a convergent-divergent type.

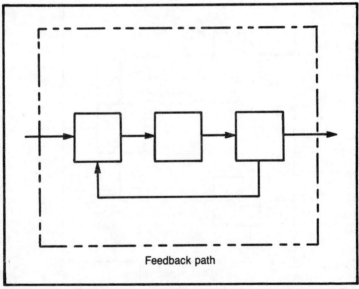

Feedback path

Fig. 3-10. The feedback path is a signal path from one circuit to a preceding circuit.

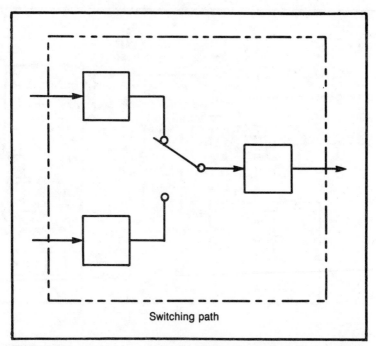

Switching path

Fig. 3-11. A circuit which has a switch for different signal paths is illustrated here.

vice in the faulty circuit. This step includes not only repairing or replacing the faulty part, but determining exactly why the failure occurred. The reason for this is that unless all faults are corrected, the trouble will possibly recur at a later date. It is also a good idea to keep records of this and all other steps so that future repairs will be made easier by referring to these written accounts of what was performed. These records may also point out consistent failures which may be due to a design error on the part of the manufacturer.

Schematic diagrams illustrate the detailed circuit arrangements of electronic parts (represented symbolically) which make up a complete circuit within a piece of equipment. These diagrams show what is inside the circuit and provide a picture of the equipment. Figure 3-12 shows an example of a schematic diagram of the receiver section of a transceiver unit. The frequency-conversion function is accomplished in a single tube. There are no separate mixer and rf oscillator circuits. Only one i-f amplifier circuit is used.

Note that the values of the circuit parts are listed. Each part is also given a reference designation for identification purposes. These diagrams will be very helpful in making tests as well as

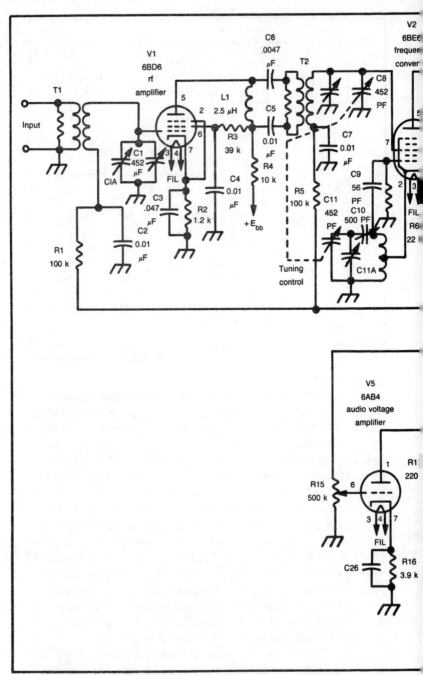

Fig. 3-12. Schematic diagram of the receiver unit of a transceiver set.

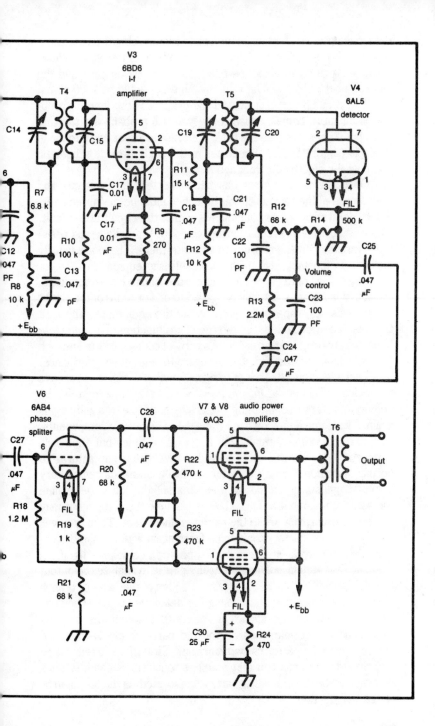

determining which branch of an isolated faulty circuit needs to be repaired. For example, to check the bias resistor of the phase-splitter tube, V6, you could place the multimeter probes on pin 7 and the junction of R_{19} and R_{21}. The value, as shown, should be 1000 ohms.

The circuit diagram discussed and pictured in this figure is probably a bit more complex than those you will probably be working with, but it does illustrate the basic principles incorporated into these types of diagrams. The appendices in this book provide schematic symbols for electrical components.

Once the faulty circuit has been isolated, the voltages and resistances of the various circuit branches must be measured in order to determine which components within the circuit are at fault. The measurement results are then compared to voltage and resistance charts or tables in order to evaluate them. These are usually included in the manuals which come with the equipment.

Regardless of the type of symptom determined in earlier steps, the actual fault can eventually be traced to one or more of the circuit parts, resistors, capacitors, etc., within the equipment. The fault may also be classified by the degree of malfunction. The complete failure or abnormal performance of a part, of course, falls in line with the explanations given earlier with regard to degraded performance.

There is a third degree of part malfunction which is not always so obvious. This is the *intermittent malfunction*. Intermittent, by definition, refers to something which alternately ceases and begins again. This same definition can be applied to electronic components. The part operates normally for a period of time, but then fails completely or operates on a degraded level, and then returns to normal operation. The cyclic nature of this malfunction can be very helpful diagnostically, although it is often still difficult to determine the actual component at fault. This is because the component may still operate normally while the circuit is under test. Thus, you will pass it by as satisfactory, only to be faced again with the same problem when the cycle of intermittent operation completes itself.

The first step in isolating the faulty part is to analyze the output signal. This will aid in making a valid selection of the parts or branch of the circuit which may be causing the defective output. The voltage, duration, and/or shape of the output waveform may be indications of possible open or shorted parts or out-of-tolerance values. This step performs two functions. First of all, it reduces to a minimum the number of test readings required. Second, it helps determine whether the faulty part, when located, is the sole source of the malfunction.

The second step in isolating a faulty part is a visual inspection of the parts and leads in the circuit. Often, this inspection will reveal burned or broken components or defective connections. Voltage measurements at transistor leads or at the pins of electron tubes in older equipment can be compared with the normal voltages listed in available voltage charts, which will provide valuable assistance in locating the source of the problem. This check will often help isolate the problem to a single branch of a circuit. A separate circuit branch is generally associated with each pin connection of the transistor or electron tube. Resistance checks at the same points are also useful in locating the trouble. Suspected parts can often be checked by a resistance measurement.

When a part is suspected of being defective, a good part may be substituted for it. Keep in mind, however, that an undetermined fault in the circuit may also damage the substituted part. Another factor to consider before performing this step is that some circuits are critical, and substituting parts (especially transistors or electron tubes) may alter the circuit parameters.

In some equipment, the circuits are specifically designed for ease of substitution. For example, the plug-in circuit module is in widespread use today. This module contains all of the necessary parts (resistors, capacitors, and inductors) for a circuit branch or even the entire circuit. Once a problem has been traced to a module, substitution is the only means of correcting it.

If you are using a voltmeter to locate the faulty component, be sure it is set to the highest scale before making measurements, and check the highest voltages first. Next, the elements having smaller voltages should be checked in descending order. Voltage, resistance, and waveform readings will seldom be identical to those specified in the performance data supplied with the equipment. The important thing at this point is to ask yourself how close to the specifications the readings should be in order to insure that the equipment is operating properly. There are many factors to consider in answering this question. The tolerance of the resistors, which greatly affects the voltage readings in a circuit, may be 20%, 10%, or 5%. In some critical circuits, precision parts are used. The tolerances marked or color-coded on the parts are, therefore, an important factor. Transistors and electron tubes have a fairly wide range of characteristics and will thus cause variations in voltage readings. The accuracy of the test instrument must also be considered. Most voltmeters have accuracies of five to ten percent, while precision meters will be more accurate.

For proper operation, critical circuits require voltage readings within the range specified by the manufacturer. However, most circuits will operate satisfactorily if the voltages are only slightly off. Important factors to consider are the symptoms and the output signal. If no output signal is produced at all, there will probably be quite a large variation of voltages in the problem area. A problem which causes the circuit to perform just out of tolerance, however, may cause only a slight change in circuit voltages.

The voltage and/or resistance checks will indicate which branch of the circuit is at fault. The next step is to isolate the problem to a particular component or components within the branch. One way to accomplish this is to move the test probe to the different points where two or more parts are joined together electrically and measure the voltage or resistance with respect to ground. Generally, however, the correct values (particularly with regard to voltage) will be difficult to determine from these points on a schematic diagram and may not be available elsewhere. Thus, this procedure should be reserved for making resistance checks to locate shorts and openings in the branch. A better method for use when checking abnormal voltage readings is a systematic check of the value of each resistor, capacitor, and/or inductor in the branch. The instruments required for these measurements include impedance bridges, Q-meters, etc.

A review of all the information obtained up to this point will help isolate other faulty parts, whether the problem is due to some isolated malfunction or to another entirely unrelated cause (multiple malfunction). In order to determine if there is a multiple malfunction, it is necessary to determine whether a fault in the one isolated component could produce all of the observed symptoms and indications. If so, you can logically conclude that it is the sole item at fault. If not, a knowledge of electronics as well as the individual piece of equipment should be used to determine what other malfunctions could also account for all of the symptoms and test data. This mental review will prove indispensable in the recognition and isolation of multiple malfunctions or any other problems caused by an isolated, faulty component. For example, consider the transistor amplifier circuit shown in Fig. 3-13. Assume that the troubleshooting procedures described here have isolated the transistor as the cause of the malfunction. It is burned out. What could cause this to happen? Excessive current can destroy the transistor by causing internal shorts or by altering the characteristics of the semiconductor material, which may be very temperature-sensitive. Thus, the problem

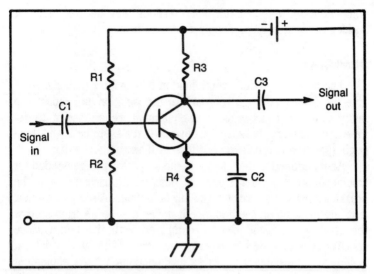

Fig. 3-13. Transistor amplifier circuit.

is reduced to a matter of determining how excessive current can be produced.

Excessive current can be caused by too great an input signal which overdrives the transistor. Such an occurrence would indicate a fault somewhere in the circuitry preceding the input connection. Power surges (intermittent, excessive outputs) from the power supply could also cause the burn-out. In fact, power supply surges are a common cause of both transistor and electron tube burn-out. Therefore, it is advisable to check for any of these conditions before placing a new transistor in the circuit. Some other malfunctions, along with their common causes include:

1. Burned-out cathode resistors caused by shorts in electron tube elements.

2. A power-supply overload caused by a short-circuit in some portion of the voltage-distribution network.

3. A burned-out transformer in the shunt feed system caused by a shorted blocking capacitor.

4. Burned-out fuses caused by power-supply surges or shorts in filtering (power) networks.

In general, a degraded component characteristic can be traced to the operating condition which caused the maximum ratings of the component to be exceeded. The condition may be temporary or accidental, or it may be deeply rooted in the circuitry itself.

SUMMARY

The six-step procedure outlined in this chapter is designed to give the reader a logical approach to the isolation and repair of a faulty component in any piece of equipment, regardless of its degree of complexity. It should be obvious that a systematic and thorough approach is much more effective than trial and error.

As mentioned earlier, some of the steps will not be needed for troubleshooting smaller pieces of equipment, particularly in the breaking down of the circuitry into functional units. Also, it is essential that any manuals on the equipment under test be on hand so that the readings obtained can be compared with the performance specifications provided by the manufacturer. Without this data, it is almost impossible to correct malfunctions where equipment performance is only slightly impaired. Even if you locate the fault, and the component has been isolated and replaced, it will still be necessary to have something to compare the new performance readings with in order to insure that the replaced component has really brought performance back up to specifications.

The trick to localizing a malfunction in any piece of equipment is to take your time, go through each step carefully and thoroughly, take notes as you proceed, and leave nothing to chance. When one component is isolated, don't stop there. Ask yourself *why* it malfunctioned in the first place so that any other problems can be corrected at the same time so as to avoid having to repeat the whole process when the isolated component fails again a short time after being replaced. This type of follow-through thinking can save many hours of labor and is well worth the extra effort.

Chapter 4

Measuring Instruments

In the field of electricity and electronics, as in other physical sciences, accurate measurements are essential. This involves two important items: numbers and units. Simple arithmetic is used in most cases, and the units are well-defined and easily understood. The standard units of current, voltage, and resistance, as well as other units of measurement, are defined by the National Bureau of Standards. At the factory, the various instruments are calibrated by comparing them with these established standards.

Electrical equipments are designed to operate at certain efficiency levels. To aid in maintaining equipment, manufacturers usually provide instruction manuals and performance specifications. These parameters should be periodically checked with the use of the appropriate measuring device to make certain that the equipment is operating according to these specifications.

In order to know the proper instrument to use for a particular piece of equipment, it is necessary to be aware of the capabilities of the various instruments. In each case, the instrument indicates the value of the quantity measured. This number is then interpreted in a way that will help the user understand the operation of the circuit. A thorough understanding of the construction, operation, and limitations of the basic types of electrical measuring instruments, coupled with the theory of circuit operation, is most essential in selecting the proper instrument and in servicing and maintaining electrical equipment.

Before going into a discussion of the basic types of meters, this chapter will discuss some principles of magnetism and electromagnetism, because some of the meters in this chapter operate by virtue of the magnetic field associated with current flow. The discussion here will not go into great detail, but rather will provide an overview of magnetic and electromagnetic properties in general.

MAGNETISM

The magnetic force surrounding a magnet is not uniform but rather is concentrated at the ends of the magnet. These ends are called the *poles* of the magnet. A magnet will always have two magnetic poles and both poles will have equal magnetic strength. The two poles are designated the north pole and the south pole. A force of attraction exists between two unlike poles, and a force of repulsion exists between two like poles.

The space surrounding a magnet where the magnetic forces act is known as the *magnetic field*. Exploration of this field yields the fact that the magnetic field is very strong at the poles and weakens as the distance from the poles increases. It is also known that the field extends from one pole of the magnet to the other, thus creating a loop around the magnet.

To further describe and work with magnetic phenomena, lines are used to represent the force existing in the area surrounding a magnet. This is shown in Fig. 4-1. These lines, known as *magnetic lines of force*, are actually only imaginary but are used to illustrate and describe the pattern of the magnetic field. The magnetic lines of force are assumed to emanate from the north pole of a magnet, pass through the surrounding space, and enter the south pole. The lines then travel inside the magnet from the south pole to the north pole, thus forming closed loops.

Although magnetic lines of force are imaginary, simple magnetic phenomena can be explained by assuming the magnetic lines have certain real properties. The main characteristics of magnetic lines of force are:

1. Magnetic lines of force are continuous and always form closed loops.
2. Magnetic lines of force never cross.
3. Magnetic lines of force travelling in the same direction repel one another. Magnetic lines travelling in opposite directions tend to attract each other and combine.

94

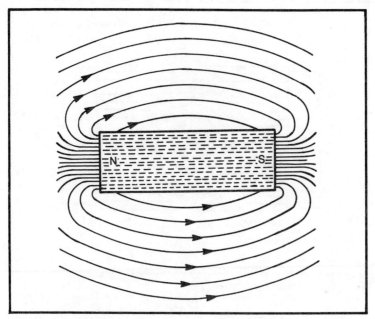

Fig. 4-1. Magnetic lines of force.

4. Magnetic lines of force tend to shorten themselves (take the shortest route).

5. Magnetic lines of force pass through all materials, both magnetic and nonmagnetic.

6. Magnetic lines of force always enter or leave a magnetic material at right angles to the surface.

The total number of magnetic lines of force leaving or entering a surface is called the *magnetic flux*. The number of flux lines per unit area is known as *flux density*. The intensity of a magnetic field is directly related to the magnetic force exerted by the field.

There is no known insulator effective against magnetic flux. Any material, when placed within a magnetic field, will be penetrated by magnetic flux. Since sensitive instruments become inaccurate when subjected to the influence of stray magnetic fields, it is necessary to protect them in some manner. Because an instrument's mechanism cannot be insulated from magnetic flux, it is necessary to redirect the passage of the flux lines. It is known that the magnetic lines of force take the path of least opposition. Therefore, if the mechanism is surrounded with a material having a high permeability, the

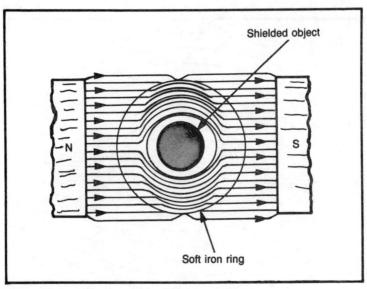

Fig. 4-2. Principles of a magnetic shield.

flux lines will take the easy path through the surrounding material. A sensitive instrument is protected by enclosing it in a soft iron case called a *magnetic shield*. This is shown in Fig. 4-2. It must be emphasized again that there is no insulator for magnetic lines of force, but by placing an instrument inside the iron shield, the lines can be redirected.

ELECTROMAGNETISM

Whenever an electron moves, it generates a magnetic field whose lines of force appear as concentric circles about the electron. A conductor (such as copper wire) contains many free electrons. The applications of an electromotive force to a conductor causes many of these free electrons to move in the same direction through the conductor. This movement of electrons constitutes a current flow. The individual magnetic fields of all the electrons moving in the same direction in the conductor will be additive, with the result that a magnetic field will exist around a conductor when a current is passing through it. Figure 4-3 illustrates the direction of flux lines around a conductor in relation to the electron flow. The relation between the direction of the magnetic lines of force around a conductor and the direction of current flow along the conductor can be determined by means of the *left-hand rule* for conductors. The rule states that if the left hand is placed so that the thumb points in the direction of electron flow, the curled fingers will point in the direction of the flux lines encircling the conductor.

Figure 4-3 also illustrates the left-hand rule for conductors. If the electron flow were reversed, the hand would have to be turned upside down with the thumb pointing toward the bottom of the page. The fingers will now indicate that with the electron flow reversed; the direction of the flux lines is also reversed.

Arrows are generally used in electric diagrams to denote the direction of current flow in a wire. Where cross sections of wire are shown, a special view of the arrow is used. A cross-sectional view of a conductor that is carrying current toward the observer is illustrated in Fig. 4-4. The direction of current is indicated by a dot, which represents the head of the arrow. A conductor that is carrying current away from the observer is illustrated in Fig. 4-5. The direction of current is indicated by a cross, which represents the tail of the arrow.

When two parallel conductors carry current in the same direction, the magnetic fields tend to encircle both conductors, drawing them together with a force of attraction, as shown in Fig.

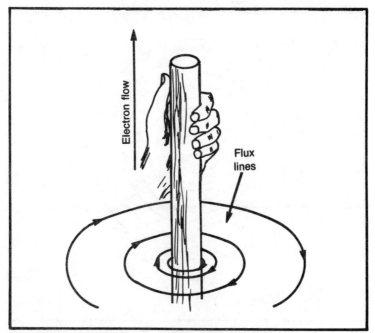

Fig. 4-3. Flux lines around a conductor in relation to current flow.

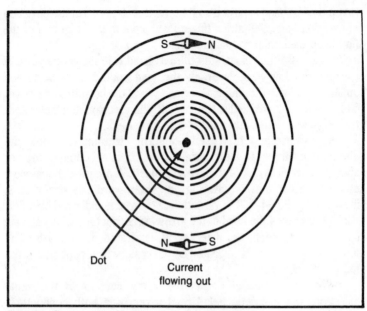

Fig. 4-4. Cross-sectional view of a conductor carrying current toward the observer.

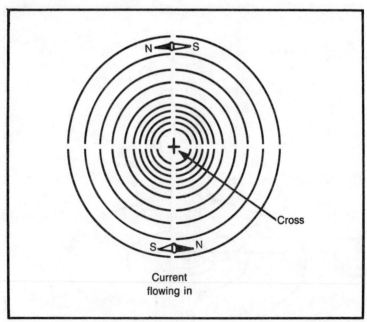

Fig. 4-5. Cross-sectional view of a conductor carrying current away from the observer.

4-6. Two parallel conductors carrying currents in opposite directions are shown in Fig. 4-7. The field around one conductor is opposite to the field around the other conductor. The resulting lines of force are crowded together in the space between the wires and tend to push the wires apart. Therefore, two parallel conductors carrying current in opposite directions repel one another.

The magnetic field around a current-carrying wire exists at all points along its length. The field consists of concentric circles in a plane perpendicular to the wire. When this straight wire is wound around a core, as shown in Fig. 4-8A, it becomes a coil and the magnetic field assumes a different shape. Part A is a partial cutaway view which shows the construction of a simple coil. Part B is a complete cross-sectional view of the same coil. The two ends of the coil are identified as points a and b. When current is passed through the coiled conductor, as indicated, the magnetic field of each turn of wire links with the fields of adjacent turns. The combined influence of all the turns produces a two-pole field similar to that of a simple bar magnet. One end of the coil will be a north pole and the other end will be a south pole.

It is already known that the direction of the magnetic field around a straight conductor depends on the direction of current flow through that conductor. Thus, a reversal of current flow through a conductor

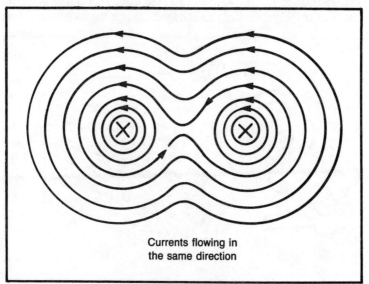

Currents flowing in
the same direction

Fig. 4-6. When two parallel conductors carry current in the same direction, the magnetic fields tend to encircle both conductors.

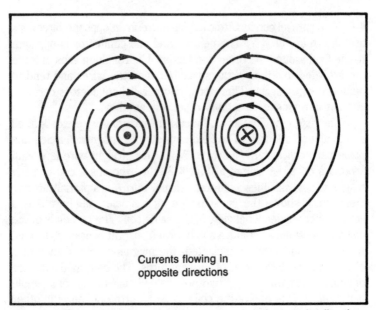

Currents flowing in
opposite directions

Fig. 4-7. When two parallel conductors carry current in opposite directions, the result looks like this.

causes a reversal in the direction of the magnetic field that is produced. It follows that a reversal of the current flow through a coil also causes a reversal of its two-pole field. This is true because the field is the product of the linkage between the individual turns

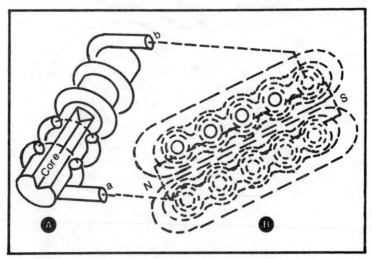

Fig. 4-8. The magnetic field around a conductory carrying current exists at all points along its length.

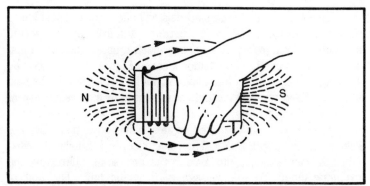

Fig. 4-9. The left-hand rule for coils.

of wire on the coil. Therefore, if the field of each turn is reversed, it follows that the total field of the coils is also reversed.

When the direction of electron flow through a coil is known, its polarity may be determined by the use of the *left-hand rule for coils*. This rule is illustrated in Fig. 4-9 and is stated as: grasping the coil in the left hand with the fingers "wrapped around" in the direction of electron flow, the thumb will point toward the north pole.

Strength of an Electromagnetic Field

The strength or intensity of a coil's field depends on a number of factors. Each will be discussed individually.

Number of turns of the conductor. When current is flowing in the same direction in two parallel conductors, the flux lines of the individual conductors will be additive. When a conductor is wound in the form of a coil, the current in adjacent turns is in the same direction. Therefore, the flux from individual turns will be additive. Increasing the number of turns will increase the number of flux lines and thereby the intensity of the coil's field.

Amount of current flow through the coil. The strength of the magnetic field around a conductor carrying a current is a function of the current. The larger the current, the stronger the magnetic field. When the conductor is wound into a coil, this relationship between current and flux still exists. Increasing the current will increase the flux around the coil turns and thereby increase the intensity of the coil's field.

Ratio of the coils length to width. In order for the field intensity to be uniform throughout the cross section of the coil, the ratio of length to width should be at least 10:1. In other words, the length of the coil should be at least 20 times the radius.

Core material. Some materials will offer more opposition to magnetic lines of force than do others. If a soft iron core is inserted in a coil, the number of flux lines is much greater than when just the air core is used. The additional lines of force are produced by the magnetization of the iron core and not by an increase in the field intensity. Field intensity can only be increased by increasing current or the number of turns.

Two terms that are used in describing core materials are *reluctance* and *permeability*. The reluctance, R, is similar to resistance in Ohm's law. Reluctance is the opposition offered by the magnetic circuit to the passage of magnetic flux. The unit of reluctance has not been named officially. However, the *rel* has been proposed and the symbol R is commonly used. One rel is the reluctance of 1 cubic centimeter of air. The reluctance of a magnetic substance varies directly as the length of the flux path and inversely as the cross-sectional area and the permeability of the substance.

Permeability, designated by the Greek letter, mu or μ, is a measure of the relative lines of force within a material as compared with air. The permeability of air is taken to be 1. Therefore, any core material having a permeability with a value greater than one will produce a greater flux density.

The strength of an electromagnetic field is expressed in *ampere turns*, because current and the number of turns are the major factors in determining the field strength of a coil. The ampere turn is equal to the product of the current through the coil and the number of coil turns.

ELECTROMAGNETIC FORCES

You already know from previous discussions that a conductor carrying a current will have a magnetic field surrounding it. You also know that if a magnet is shaped in such a way that its poles are close together, a magnetic field will exist between them. The magnetic field between the north and south poles of a magnet is shown in Fig. 4-10. The lines of force comprising the field extend from the north pole to the south pole. A cross section of a current-carrying conductor is shown in Fig. 4-11. Here, the plus sign in the wire indicates that the electron flow is away from the observer. The direction the flux loops around the wire is counterclockwise as shown. This follows the left-hand rule for conductors.

If the conductor (carrying the electron flow away from the observer) is placed between the poles of the magnetic as shown in Fig. 4-12, both fields will be distorted. Above the wire the field is

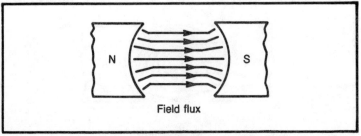

Fig. 4-10. The magnetic field between the north and south pole of a magnet.

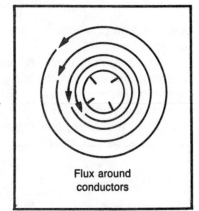

Fig. 4-11. Cross section of a current-carrying conductor.

Flux around conductors

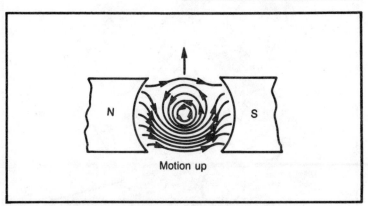

Motion up

Fig. 4-12. A conductor can be used to distort the magnetic field.

weakened, and the conductor tends to move upward. The force exerted upward depends on the strength of the field between the poles and on the strength of the current flowing through the wire.

If the current through the conductor is reversed, as shown in Fig. 4-13, the direction of the flux around the wire is reversed. The

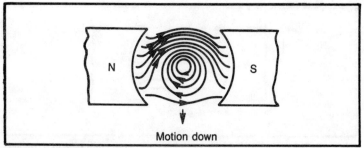

Fig. 4-13. Effective current reversal through a conductor placed in a magnetic field.

field below the conductor is now weakened and the conductor tends to move downward.

A convenient method of determining the direction of motion of a current-carrying conductor in a magnetic field is by the use of the *right-hand motor rule*, which is illustrated in Fig. 4-14. The right-hand motor rule says that to find the direction of motion of a conductor, the thumb, first finger, and second finger of the right

Fig. 4-14. The right-hand motor rule.

hand are extended at right angles to each other. The first finger is pointed in the direction of the flux (toward the south pole), and the second finger is pointed in the direction of electron flow in the conductor. The thumb then points in the direction of motion of the conductor with respect to the field. The conductor, the field, and the force are mutually perpendicular to each other.

The force acting on a current-carrying conductor in a magnetic field is directly proportional to the field strength of the magnet, the active length of the conductor, and the intensity of the electron flow through it.

TYPES OF METER MOVEMENTS

The stationary, permanent-magnet, moving-coil meter is used in many measuring instruments for servicing electrical equipment. The basic movement consists of a stationary permanent magnet and a movable coil. When current flows through the coil, the resulting magnetic field reacts with the magnetic field of the permanent magnet and causes the coil to rotate. The greater the current through the coil, the stronger the magnetic field. The stronger this field is, the greater will be the rotation of the coil.

In the D'Arsonval-type meter, the length of the conductor is fixed and the strength of the field between the poles of the magnet is fixed. Therefore, any change in I causes a proportionate change in the force acting on the coil. The principle of the D'Arsonval movement may be more clearly shown by the use of the simplified diagram in Fig. 4-15, which illustrates the movement commonly used in dc instruments. Here, only one turn of wire is shown. However, in an actual meter movement, many turns of fine wire would be used, each turn adding more effective length to the coil. The coil is wound on an aluminum frame or bobbin, to which the pointer is attached. Oppositely wound hairsprings (one of which is shown in Fig. 4-15) are also attached to the bobbin, one at either end. The circuit to the coil is completed through the hairsprings. In addition to serving as conductors, the hairsprings serve as the restoring force that returns the pointer to the zero position when no current flows.

It is already known that the deflecting force is proportional to the current flowing in the coil. The deflecting force tends to rotate the coil against the restraining force of the hairspring. The angle of rotation is proportional to the deflecting force. When the deflecting force and the restraining force are equal, the coil and the pointer cease to move. Since the deflecting force is proportional to the current in the coil and the angle of rotation is proportional to the

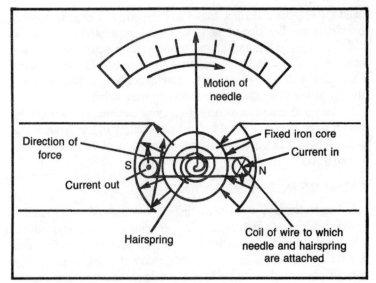

Fig. 4-15. The basic D'Arsonval movement.

deflecting force, the angle of rotation is proportional to the current through the coil. When current ceases to flow in the coil, the driving force ceases, and the restoring force of the springs returns the pointer to the zero position.

If the current through the single turn of wire is in the direction indicated (away from the observer on the right-hand side and toward the observer on the left-hand side), the direction of force, by the application of the right-hand motor rule, is upward on the left-hand side and downward on the right-hand side. The direction of motion of the coil and pointer is clockwise. If the current is reversed in the wire, the direction of motion of the coil and pointer is reversed.

A detailed view of the basic D'Arsonval movement as commonly employed in ammeters and voltmeters is shown in Fig. 4-16. The iron core is rigidly supported between the pole pieces and serves to concentrate the flux in the narrow space between the iron core and the pole piece; in other words, in the space through which the coil and the bobbin moves. Current flows into one hairspring, through the coil, and out of the other hairspring. The restoring force of the spiral springs returns the pointer to the normal or zero position when the current through the coil is interrupted. Conductors connect the hairsprings with the outside terminals of the meter. If the instrument is not damped; that is, if some type of loss is not introduced to absorb the energy of the moving element, the pointer will oscillate for a long time about its final position before coming to rest. This action makes it nearly impossible to obtain a reading, and some form of damping is necessary to make the meter practicable. Damping is accomplished in many D'Arsonval movements by means of the motion of the aluminum bobbin upon which the coil is wound. As the bobbin oscillates in the magnetic field, an EMF is induced in it because it cuts through the lines of force. Therefore, induced currents flow in the bobbin in such a direction as to oppose the motion, and the bobbin quickly comes to rest in the final position after going beyond it only once.

In addition to factors such as increasing the flux density in the air gap, the overall sensitivity of the meter can be increased by the

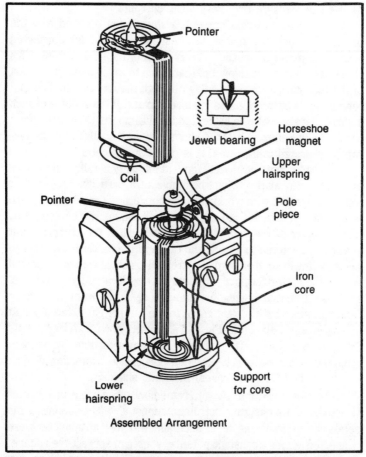

Fig. 4-16. The D'Arsonval movement as employed in ammeters and voltmeters.

use of a lightweight rotating assembly (bobbin, coil, and pointer) and by the use of jewel bearings. Note in Fig. 4-16 that the pole pieces have curved faces. The advantage of this type of construction can be seen if it is remembered that lines of force enter and leave a magnetic surface at right angles. The curved surfaces produce a uniform magnetic field in the air gap at right angles to the coil, regardless of the coil's angular position. This type of construction makes possible a more linear scale than if the pole faces were flat.

THE DC AMMETER

The small size of the wire in the ammeter's movable coil places limits on the current that may be passed through the coil. Consequently, the basic D'Arsonval movement discussed so far may be used to measure only very small currents.

To measure a larger current, a *shunt* must be used with the meter. A shunt is a heavy, low-resistance conductor connected across the meter terminals to carry most of the load current. This shunt has the correct amount of resistance to cause only a small part of the total circuit current to flow through the meter coil. The meter current is proportional to the load current. If the shunt is of such a value that the meter is calibrated in milliamperes, the instrument is called a *milliammeter*. If the shunt is of such a value that the meter is calibrated in amperes, it is called an *ammeter*.

A single type of meter movement is generally used in all ammeters, no matter what the range of a particular meter. For example, meters with working ranges of zero to ten amperes, zero to five amperes, or zero to one ampere all use the same movement. The designer of the ammeter simply calculates the correct shunt resistance required to extend the range of the meter movement to measure any desired amount of current. This shunt is then connected across the meter terminals. Shunts may be located inside the meter case (internal shunt) or somewhere away from the meter (external shunt) with leads going to the meter. An external shunt arrangement is shown in Fig. 4-17. Some typical external shunts are shown in Fig. 4-18. Figure 4-19 shows a meter movement mounted in a case which provides portability, protection against breakage, and (in some cases) magnetic shielding.

The most important thing to remember when using an ammeter is that current measuring instruments must *always* be connected in series with a circuit and *never* in parallel with it. If an ammeter were connected across a constant-potential source of appreciable voltage, the shunt would become a short circuit and the meter would burn out.

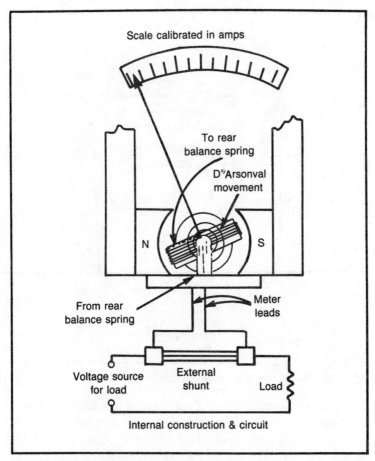

Fig. 4-17. An external shunt arrangement in a dc ammeter.

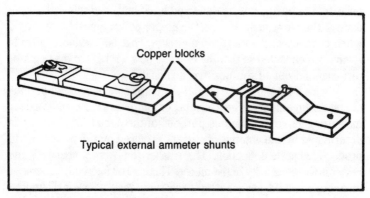

Fig. 4-18. Typical ammeter shunts.

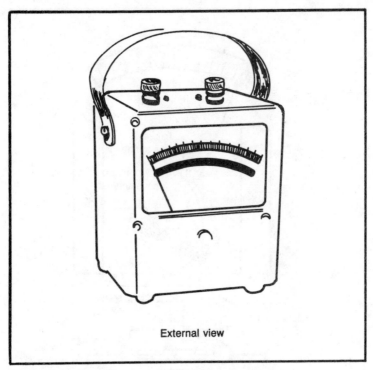

External view

Fig. 4-19. A meter movement mounted in a protective case.

If the approximate value of current in a circuit is not known, it is best to start with the highest range of the ammeter and switch to progressively lower ranges until a suitable reading is obtained. Most ammeter needles indicate the magnitude of the current by being deflected from left to right. If the meter is connected with reversed polarity, the needle will be deflected backward, and this action may damage the movement. Hence, the proper polarity must be observed when connecting the meter in a circuit. That is, the meter should always be connected so that the electron flow will be into the negative terminal and out of the positive terminal.

Figure 4-20 shows the proper ammeter connections for various circuit arrangements. The ammeter or ammeters are connected for measuring current in various portions of the circuits. Sensitivity is determined by the amount of current required by the meter coil to produce full scale deflection. The smaller the current required, the better is the sensitivity of the meter. Thus, a 100 microampere meter movement would have better sensitivity than a one milliampere movement.

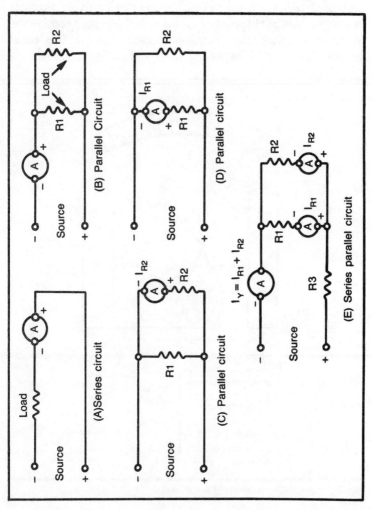

Fig. 4-20. Ammeter connections for various types of applications.

111

THE DC VOLTMETER

The 100-microampere D'Arsonval meter used as the basic meter for the ammeter may also be used to measure voltage if a high resistance is placed in series with the moving coil of the meter. For low-range instruments, this resistance is mounted inside the case with the D'Arsonval movement and typically consists of resistance wire having a low temperature coefficient and wound either on spools or card frames. For higher voltage ranges, the series resistance may be connected externally. When this is done, the unit containing the resistance is commonly called a *multiplier*.

A simplified diagram of a voltmeter is shown in Fig. 4-21. The resistance coils are treated in such a way that a minimum amount of moisture will be absorbed by the insulation. Moisture reduces the insulation resistance and increases leakage currents, that cause incorrect readings. Leakage currents through the insulation increase

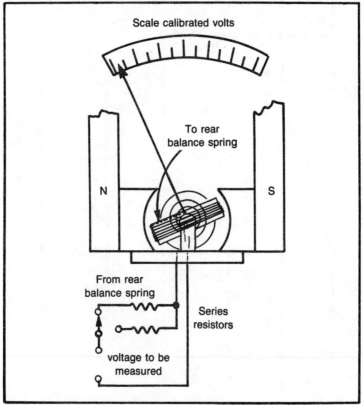

Fig. 4-21. Simplified diagram of a voltmeter.

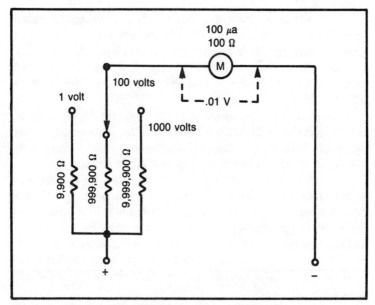

Fig. 4-22. Schematic drawing of a multirange voltmeter.

with the length of the resistance wire and limits the maximum measurable voltage.

The value of the necessary series resistance is determined by the current required for full-scale deflection of the meter and by the range of voltage to be measured. Because the current through the meter is directly proportional to the applied voltage, the scale can be calibrated directly in volts for a fixed series resistance.

Multirange voltmeters utilize one meter movement with the required resistance connected in series with the meter by a convenient switching arrangement. A schematic diagram of a multirange voltmeter with three ranges is shown in Fig. 4-22.

A very important point to keep in mind when using a voltmeter is that any voltage-measuring instrument is connected across or in parallel with a circuit. If the approximate value of the voltage to be measured is not known, it is best to start with the highest range of the voltmeter and progressively lower the range until a reading is obtained.

In many cases, the voltmeter is not a center-zero indicating instrument. Thus, it is necessary to observe the proper polarity when connecting the instrument to a circuit, as with the dc ammeter. The voltmeter is connected so that electrons will flow into the negative terminal and out of the positive terminal of the meter.

The function of a voltmeter is to indicate the potential difference between two points in a circuit. When the voltmeter is connected across a circuit, it shunts the circuit. If the voltmeter has low resistance, it will draw an appreciable amount of current. The effective resistance of the circuit will be lowered and the voltage readings will consequently be lowered.

When voltage measurements are made in high-resistance circuits, it is necessary to use a high-resistance voltmeter to prevent the shunting action of the meter. The effect is less noticeable in low-resistance circuits because the shunting effect is less.

The sensitivity of a voltmeter is given in ohms per volt and may be determined by dividing the resistance of the meter plus the series resistance by the full-scale reading in volts. The accuracy of the meter is generally expressed in percentages. For example, a meter that has an accuracy of one percent will indicate a value that is within one percent of the correct value. This means that if the correct value is 100 units, the meter indication may be anywhere from 99 to 101 units.

THE OHMMETER

The ohmmeter is widely used to measure resistance and check the continuity of electrical circuits and devices. Its range usually extends to only a few megohms. This instrument consists of a dc milliammeter, which was mentioned earlier in this chapter, plus two additional features. One is a dc voltage source usually a three-volt battery. The other is one or more resistors, one of which is variable. A simpler ohmmeter circuit is shown in Fig. 4-23.

The ohmmeter's pointer deflection is controlled by the amount of battery current passing through the moving coil. Before measuring the resistance of an unknown resistor or electrical circuit, the ohmmeter must be calibrated. If the value of resistance to be measured can be estimated within reasonable limits, a range is selected which will give approximately half-scale deflection when this resistance is inserted between the probes. If the resistance is unknown, the selector switch is set on the highest range. Whatever range is selected, the meter must be calibrated to read zero before the unknown resistance is measured.

Calibration is accomplished by first shorting the test leads together, as shown in Fig. 4-23. With the test leads shorted, there will be a complete series circuit consisting of the 3 V source, the resistance of the meter coil (R_m), the resistance of the zero-adjust potentiometer, and the series multiplying resistor (R_s). Current will

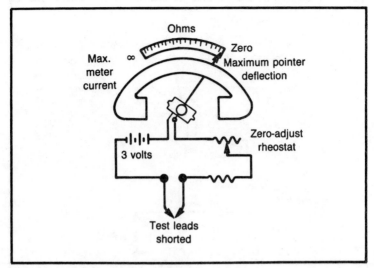

Fig. 4-23. A basic ohmmeter circuit.

flow and the pointer will be deflected. The zero division of the meter under discussion and in many commercial ohmmeters is located at the extreme right of the scale. With the test leads shorted, the zero-adjust potentiometer is set so that the pointer rests on zero. Therefore, full-scale deflection indicates zero resistance between the test leads. If the range is changed, the meter must be zeroed again to obtain an accurate reading. When the test leads of an ohmmeter are separated, the pointer of the meter will return to the left side of the scale. This is due to the interruption of current and the spring tension acting on the movable coil assembly. This reading indicates infinity.

After the ohmmeter is adjusted for zero reading, it is ready to be connected in a circuit to measure resistance. A typical circuit and ohmmeter arrangement is shown in Fig. 4-24. The power switch of the circuit to be measured should always be in the OFF position. This prevents the source voltage from being applied across the meter, which could damage the meter movement.

The test leads of the ohmmeter are connected across (in series with) the circuit to be measured. This causes the current produced by the meter's 3 V battery to flow through the circuit being tested. Assume that the meter test leads are connected at points a and b of Fig. 4-24. The amount of current that flows through the meter coil will depend on the resistance of resistors R_1 and R_2, plus the resistance of the meter coil and zero-adjust rheostat. Because the

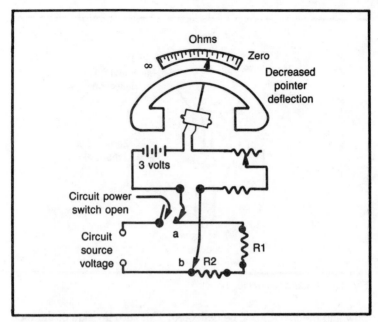

Fig. 4-24. A typical circuit and ohmmeter arrangement.

meter has been preadjusted (zeroed), the amount of coil movement now depends solely on the resistance of R_1 and R_2. The inclusion of R_1 and R_2 raises the total series resistance, decreases the current, and thus decreases the pointer deflection. The pointer will now indicate the combined resistance of R_1, R_2. If R_1, R_2, or both are replaced with a resistor or resistors having a larger resistance, the current flow in the moving coil of the meter would be decreased even more. The deflection would also be less, and the scale indication would read an even higher circuit resistance. The movement of the moving coil is proportional to the amount of current. The scale reading of the meter in ohms is inversely proportional to current flow in the moving coil.

The amount of circuit resistance to be measured may vary over a wide range. In some cases, it may be only a few ohms, while in others, it may be as great as 1,000,000 ohms. To enable the meter to indicate any value being measured with the least amount of inaccuracy, scale multiplication features are incorporated in most ohmmeters. For example, a typical meter will have four test lead jacks marked COMMON, R × 1, R × 10, and R × 100. The jack marked COMMON is connected internally through the battery to one side of the moving coil. The jacks marked R × 1, R × 10, and

R × 100 are connected to three different size resistors located within the ohmmeter. This is illustrated in Fig. 4-25.

Some ohmmeters are equipped with a selector switch for selecting the multiplication scale desired. In this type of instrument, only two test lead jacks are provided. The range to be used in measuring any unknown resistance (R_x in Fig. 4-25) depends on the approximate ohmic value of the unknown resistance. For instance, assume the ohmmeter scale in Fig. 4-25 is calibrated in divisions from zero to 1000. If R_x is greater than 1000 ohms and the R × 1 range is being used, the ohmmeter cannot measure it. This occurs because the combined series resistance of resistor R × 1 and R_x is too great to produce sufficient battery current to deflect the pointer away from infinity. The test lead would have to be plugged into the next range, R × 10. With this done, assume the pointer deflects to indicate 375 ohms. This would indicate that R_x has 375 × 10 = 3,750 ohms resistance. The change of range caused the deflection because resistor R × 10 has only one-tenth the resistance of resistor R × 1. Thus, selecting the smaller series resistance permits

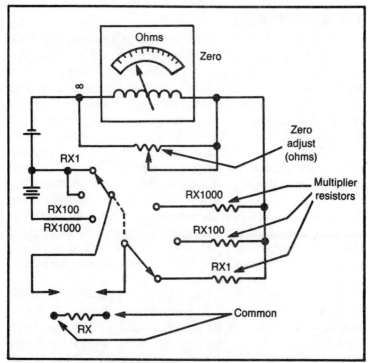

Fig. 4-25. Scale multiplication in an ohmmeter.

sufficient current to cause pointer deflection. If the R × 100 range were used to measure the 3750-ohm resistor, the pointer would deflect still further to the 37.5-ohm position. This increased deflection would occur because resistor R × 100 has only one-tenth the resistance of resistor R × 10.

The foregoing circuit arrangement allows the same amount of current to flow through the moving coil whether the meter measures 10,000 ohms on the R × 1 scale, 100,000 ohms on the R × 10 scale, or 1,000,000 ohms on the R × 100 scale.

It always takes the same amount of current to deflect the pointer to a certain position on the scale (midscale position, for example), regardless of the multiplication factor being used. Since the multiplier resistors are of different values, it is necessary to *always* zero adjust the meter for each multiplication factor selected. You should select the multiplication factor that will result in the pointer coming to rest as near as possible to the midpoint of the scale. This allows you to read the resistance more accurately, because the scale readings are more easily interpreted at or near midpoint.

THE MEGOHMMETER

An ordinary ohmmeter cannot be used for measuring resistance of many millions of ohms, such as in conductor insulation. To adequately test for insulation breakdown, it is necessary to use a much higher voltage than is furnished by an ohmmeter's battery. This potential is placed between the conductor and the outside surface of the insulation. An instrument called a *megohmmeter*, or "megger" as it is sometimes called, is used for these types of tests. Figure 4-26 shows a typical megohmmeter. This is a portable instrument that consists of two primary elements. The first is a hand-driven dc generator, G, that supplies the necessary voltage. The second is the instrument portion, which indicates the value of the resistance being measured. The instrument portion is of the opposed-coil type. Coils a and b are mounted on the movable member, c, that have a fixed angular relationship to each other and are free to turn as a unit in a magnetic field. Coil b tends to move the pointer counterclockwise and coil a clockwise.

Coil a is connected in series with R_3 and the unknown resistance, R_x, to be measured. The combination of coil a, R_3 and R_x forms a direct series path between the + and − brushes of the dc generator. Coil b is connected in series with R_2 and this combination is also connected across the generator. There are no restraining springs on the movable member of the instrument portion

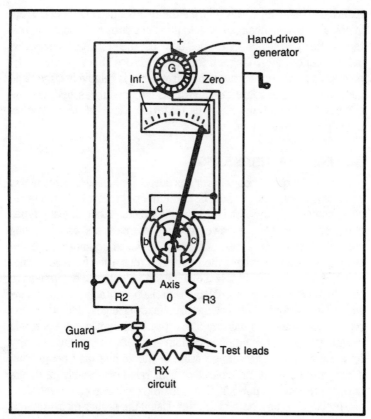

Fig. 4-26. A megohmmeter circuit.

of the meter. Therefore, when the generator is not operated, the pointer floats freely and may come to rest at any position on the scale.

The guard ring intercepts leakage current. Any leakage currents intercepted are shunted to the negative side of the generator. They do not flow through coil a; therefore, they do not affect the meter reading.

If the test leads are open-circuited, no current flows in coil a. However, current flows internally through coil b and deflects the pointer to infinity, which indicates a resistance too large to measure. When a resistance such as R_x is connected between the test leads, current also glows in coil a, tending to move the pointer clockwise. At the same time, coil b still tends to move the pointer counterclockwise. Therefore, the moving element composed of both coils and the pointer comes to rest at the position at which the two forces are balanced. This position depends upon the value of the external resistance, which controls the relative magnitude of current in coil *a*. Because changes in voltage affect both coil *a* and coil *b* in the same proportion, the position of the moving system is independent of the voltage. If the test leads are short-circuited, the pointer rests at zero because the current in *a* is relatively large. The instrument is not injured under these circumstances because the current is limited by R_3. The external view of one type of megohmmeter is shown in Fig. 4-27.

AC LINE MEASUREMENTS

While the discussion thus far has dealt with measurements within electronic equipment, it must be remembered that most circuits receive power from the standard 115-volt ac line. Although many types of equipment are battery-powered, many more depend on alternating current. Also, many types of battery-powered equipment are designed to operate and have their batteries recharged from an ac line.

It is very important when dealing with electronic equipment to be able to determine any problems with the power source. Conditions such as low voltage, high voltage, frequency changes, and other line-related problems can directly affect the operation of an electronic circuit. It is not uncommon in certain areas for the voltage to vary to some degree. The voltage value of the ac line may range from a low peak of about 105 volts to a high value of possibly 120 volts during intermittent periods. This lack of regulation can cause serious problems in an electronic circuit. Using the proper measurement

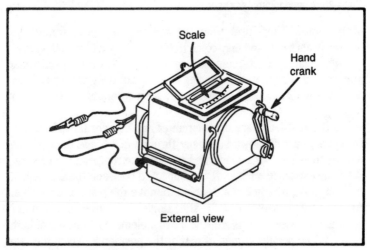

Fig. 4-27. A typical megohmmeter.

instruments, it is possible to detect these variations and provide an alternate source of power when it is determined that these fluctuations may damage the device.

THE VOLTMETER

The previous discussion on voltmeters was aimed specifically at making dc measurements. The same instrument can be used to measure ac line voltage as well. To do this, the instrument is connected across the terminals of the ac line, usually at a receptacle. This is accomplished by placing the probes into each slot in the socket. Make sure that the scale of the meter is adequate for reading the amount of voltage selected. For example, a 150-volt scale would suffice for measuring standard house-current values, which are between 110 and 120 volts. A 300-volt scale would be necessary for measuring the voltage at the service entrance, which is normally between 220 and 240 volts. Voltmeters used by electricians are well suited for ac line measurements. These meters are not enclosed but instead are open and have a straight rather than a curved scale that is usually calibrated from 100 to 600 volts. A sliding indicator is driven to a point on the scale which is the equivalent of the voltage value. These types of meters will usually measure ac or dc voltage without any switching or adjustments. They are referred to as the *electrodynamometer* type, because the field produced to drive the indicator to its proper point on the scale is obtained by the use of an electromagnet. This is powered from the voltage source rather than by a permanent magnet, which puts it into the category of the D'Arsonval type of meter discussed earlier.

THE AC METER

You already know that to take current measurements with an ammeter, it is necessary to break the circuit, place the ammeter in series, and observe the meter reading. While this may work well with electronic circuits in which most of the circuitry is located in one small area, it is sometimes very inconvenient and difficult when measuring line current. For most ac line applications, the clamp-on ac ammeter shown in Fig. 4-28 is used. With this type of instrument, it is unnecessary to place the meter in series, nor are any physical connections to the line required.

To operate this meter, the jaws of the device are opened and snapped around the power cord of the equipment under test. These jaws house a coil which is inductively coupled to the ac line. The

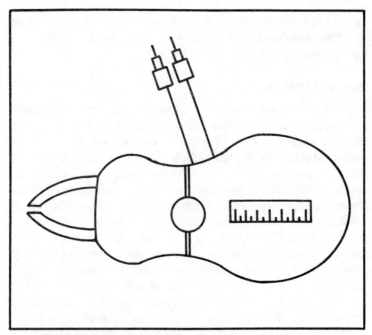

Fig. 4-28. Clamp-on ac ammeter.

current which is being drawn by the equipment at the end of the power cord or cable will be indicated on the meter. The readings from this type of meter are not as accurate as those obtained with an in-line type of meter, but, they are accurate enough for most uses where the current drain from the ac line is above one ampere.

Figure 4-29 shows how this instrument is clamped to the ac line to take measurements. This measuring is taken by actually forming an inductive circuit. The line carrying the current acts as the primary of the transformer, while the coil in the ammeter acts as the secondary. Voltage from the primary is transferred to the secondary and the meter simply reads the output voltage of the secondary winding. To extend the range of the ac ammeter, a transformer which is also of the clamp-on variety can be added to the circuit. Figure 4-30 shows such an arrangement. This can be thought of as a step-down transformer and can be clamped to a conductor or power cable which carries a current higher than that which can be measured by the ac ammeter. This transformer takes a higher voltage at the primary and steps it down to a lower voltage at the secondary. When the ac ammeter is clamped to the transformer, it reads a voltage which is reduced by a specific figure. A 3:1 transformer, for example,

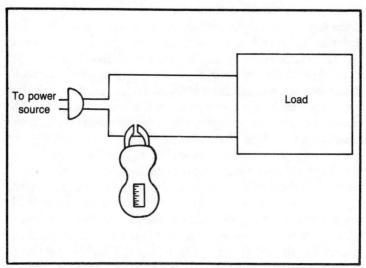

Fig. 4-29. Connection of an ac ammeter to one conductor of a power line.

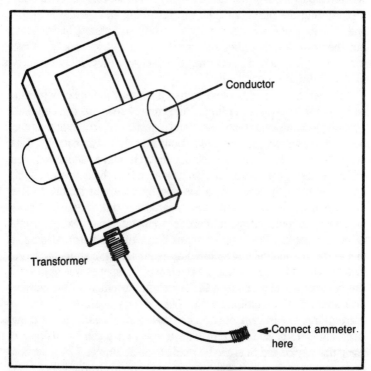

Fig. 4-30. A transformer may be used to extend the range of a clamp-on ac ammeter.

123

would produce a reading of one-third of the actual current present in the line. Thus, the meter reading taken with this circuit will be multiplied by three to arrive at the proper current measurement.

The ac ammeter may also be used to troubleshoot an electronic circuit without the necessity of gaining access to the circuit itself. If the average current demand from the ac line during normal operating conditions is known, a sharp increase or decrease in the current demand will indicate problems in the circuit. If faulty operation is suspected, a quick ac current measurement may indicate the general area of the problem as well as its possible cause.

When it is desirable to have a continuous record of the current flowing in an electrical circuit, a *recording ammeter* is often used. A recording ammeter is an instrument which constantly records the current flow in a specific circuit and relays its findings to a paper graph. This graph is continuously moving at a very slow speed and is usually graduated into time of day or night. When the chart is set in motion, it will indicate the current drain on the graph. Thus, a written indication is provided of exactly what the current conditions were during the period of time the recording ammeter was in the line. The type of charts used will vary from instrument to instrument, but they are usually either the circular or the strip variety. While the strip variety is more accurate, the circular variety is more easily filed and is more convenient.

Sometimes it is desirable to use a current multiplier with the ac ammeter to extend its range. High current readings can be taken by using a clamp-on transformer, and a device called a *multipler* can be used to multiply the value of various amounts of ac current. Figure 4-31 shows such a device. Notice that it has three separate holes which are designed to provide clamping access to the ammeter. Each hole multiplies the current in the line by a certain factor. In the discussion of the 3:1 transformer earlier in this chapter, it was necessary to multiply the ammeter reading by 3 to obtain the proper measurement. When using a multiplier, the ammeter reading is divided by the multiplying factor. For example, if the ammeter were inserted in the times 2 slot and a reading of 4 amperes was obtained, this reading would be 2 times the actual measurement of the circuit, or 2 amperes. The ammeter multiplier is very useful in electronic current-drain measurements because these circuits often do not draw large amounts of line current. These low values can be multiplied using this device and thus can be read on the ac ammeter on a portion of its meter scale which will provide more accurate readings. Low values of current will often be indicated in the lower quarter of the

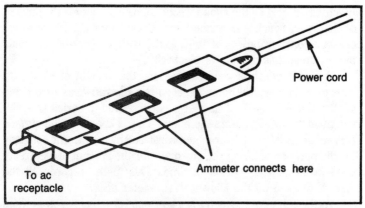

Fig. 4-31. A multiplier can be used with an ac ammeter to multiply current values.

ammeter scale in many cases. This portion of the scale does not produce the most accurate readings. By using the multiplier, these low readings are brought up to a level which can be read in the upper half of the meter scale where greater accuracy can be obtained.

To obtain accurate ac ammeter readings, make certain that the frequency of the ac current being measured is 60 Hz or the correct frequency for which the meter was designed. Ac ammeters are frequency sensitive devices and will not operate properly if the frequency differs greatly from what it was designed for. Standard ac house current is set at a fixed frequency of 60 Hz, and most ac ammeters are designed for this frequency. A few, however, may be designed for 400-Hz circuits and will not give accurate 60-Hz measurements. If a transformer is used in the electronic equipment which is connected to the ac line, take the measurements at a point on the power line which is as far from the transformer as possible. This will prevent stray coupling from the transformer circuit. Make certain that the jaws of the ammeter are clamped around only the line cord connected to the equipment. Any stray conductors will cause erratic and inaccurate meter readings. Do not use an ac ammeter to take current readings on high-voltage circuits without using high-voltage protection equipment. The methods described in this chapter for ac line measurements are meant to be used only with standard 110- to 120- volt 60-Hz circuits.

THE KILOWATT-HOUR METER

Electrical power in a circuit is measured in watts, which is the equivalent of multiplying the ac line voltage by the current in amperes.

Electricity is sold by the *kilowatt hour*. A kilowatt hour is the amount of electricity which is consumed during one hour by a device which draws a constant power of 1000 watts. Ten kilowatt-hours would be ten times this amount of electricity.

A meter which is used to measure the amount of electricity consumed in a typical home is called the *kilowatt-hour meter* and is shown in Fig. 4-32. This meter is connected in series with the electrical line which enters the home. This is known as the *surface connection*. A kilowatt-hour meter is installed by the electric company for the purpose of billing the customer for the amount of electricity used in a specified period of time. This meter records all the electricity used on the kilowatt-hour meter dials.

At the center of the meter is a horizontal disc that revolves when electricity passes through the coils of the meter. This disc can be thought of as the wheel of an electric motor and is powered by the current drain in the home. The speed of this disc is proportional to the rate power consumption. As the disc turns, a gear train moves the pointers on the dials to record the amount of power in kilowatt hours. At the end of the billing period, representatives from the electric company read each of the meters, and the homeowner receives a bill for the amount of electricity that is recorded.

While this type of meter is rarely used in practical applications by the electronics technician or the home experimenter, it does utilize the same basic principles by which all electricity is measured. Another type of meter which is used for these types of measurements is the *watt-hour meter*. Although the watt-hour meter is not a standard instrument, it is worth mentioning. The watt-hour meter

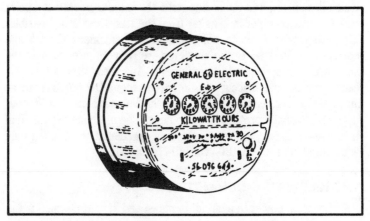

Fig. 4-32. A kilowatt-hour meter.

functions in basically the same way as the kilowatt-hour meter, except that it reads the smaller amounts of current. A watt-hour is equivalent to the amount of electricity used during one hour by a device which draws one watt of power.

SUMMARY

In order to best use electronic test instruments, it is necessary to know a little about the electronic components from which they are built and also to understand the principles under which readings are taken. Multimeters, which will be discussed in the next chapter, are actually a combination of many of the single-purpose devices discussed here. The basic principle of electromagnetism under which these devices operate is now very old indeed, but modern design capabilities and closer manufacturing tolerances have taken full advantage of this principle to allow it to be used for highly critical measurements of electronic values.

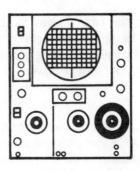

Chapter 5

The Multimeter—The Universal Test Instrument

The *multimeter* is a multipurpose instrument combining the features of an ammeter, voltmeter, and ohmmeter in one instrument. One meter movement is used for all functions, with the face of the instrument having separate graduated scales for each of the various functions. Figure 5-1 shows a typical multimeter, which is neither expensive nor difficult to operate. This instrument is used more than any other and a day rarely goes by that its use is not dictated when experimenting with or servicing electronic circuits. Unfortunately, many people do not take full advantage of the multimeter because they don't fully understand all of its capabilities.

The multimeter is often referred to as an ohmmeter, ammeter, or db meter. All of these terms are technically correct because even the most inexpensive models can usually perform measurements in all of these categories. A multimeter is usually self-contained, having its own power source, but sometimes it will require 110-volt house current. Some models are called *vacuum tube voltmeters* (VTVM) and use vacuum tubes to provide very high accuracy in voltage measurements. Most of these, however, have long been replaced by solid-state equivalents which provide even better accuracy.

Some multimeters are available with a standard meter readout of the values being measured, while others are now available in digital readout forms which provide easy-to-read digital displays of electronic values. A multimeter will usually have two test probes that are used to measure values such as voltage and resistance in addition to

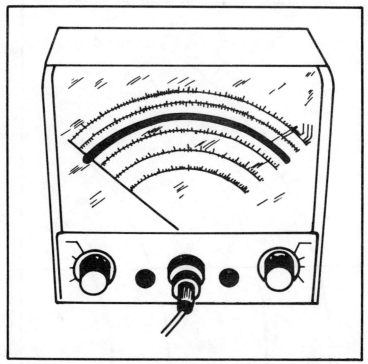

Fig. 5-1. A typical multimeter that is often seen on the test bench.

amperage without inserting these probes in different sockets. A mode switch on the multimeter automatically switches the probe leads into the correct portion of the multimeter circuit. Some of the more expensive multimeters will contain three probes. One is a ground connection when measuring voltage circuits, while the other is the positive probe for dc circuits. When measuring ac circuits, the dc probe contains a switch which is thrown to the ac position and the same probe is used for the latter circuit. The ground lead is also used in conjunction with the third lead to measure resistance. In this mode, the voltage probe is not used. Sometimes, for high voltage or high current measurements, a separate probe slot is provided. In this case, the positive probe is removed from its normal slot and inserted into the other in order to obtain the special readings. Most multimeters measure from zero to several million ohms and from zero to about a thousand volts ac or dc, with a special probe slot for use up to about 5,000 volts.

Figure 5-2 shows a typical multimeter which provides several different scales on one meter faceplate. One scale is used to measure

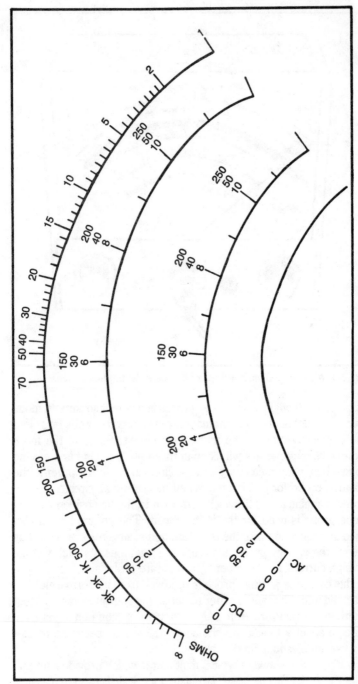

Fig. 5-2. The meter faceplate of most multimeters contains many different scales.

130

ohms and is usually calibrated from zero to infinity. When the resistance range switch is in the ×1 position, the meter will be read according to the printed scale. In the ×10 position, all readings are multiplied by 10. The same is true of the voltage scale, which is used to measure voltages in several different ranges by simply multiplying by the factor indicated on the value switch.

The multimeter is a very accurate device if a good instrument is purchased. The accuracy is determined by the quality and tolerance of the internal resistors used. The cost of a high-quality instrument is mainly in the precision resistors. An inexpensive unit will contain resistors which are not as close to their stated values due to larger tolerances. These instruments will generally give good measurements, but they can vary as much as 10% to 15%, and in some cases even 20%, from the actual value being measured.

Almost every electronics shop or test bench will contain a multimeter in some form or other. When a voltage is present in a circuit and its value needs to be known, it is the multimeter which is normally used. The same is true when a resistor needs to be checked, a battery is thought to be weak, or the current drain of a circuit needs to be known. The multimeter probes are placed into the electronic circuit in many different ways, depending on the electronic condition being read. For instance, if dc voltage is being read, the probes will be placed with the negative one at circuit ground and the positive probe on the positive side of the circuit. The selector has already been thrown into the dc voltage mode and a range selected which should be somewhere in the area of the anticipated voltage. If the ballpark value of the unknown voltage cannot be determined, the multimeter range switch is set to measure the highest dc voltage possible with this instrument. When the probes are properly placed and the circuit activated, the range switch will then be adjusted downward until a proper reading is indicated. This is a safety precaution and protects the instrument from damage. If the probes of the multimeter are being used to measure a circuit which contains a potential of 500 volts dc and the range switch is in the 100-volt position, as soon as the circuit is activated, the indicator will go completely off scale; that is, past the 100-volt position, which is the maximum it is designed to read with at this setting. If a standard meter indicator is used, the meter needle may be severely bent. The internal mechanism of the meter may be destroyed, and some of the precision resistors within the multimeter circuit may be destroyed also. This is why it is necessary to have a rough idea of the range of values you are measuring. If you are in doubt, set the range switch to the highest value available and then work downward until the range approaches the value you are measuring. This method applies only to measuring voltage or current with the multimeter. When measuring ohms, if the meter goes off scale, no serious damage will result.

Most multimeters have one or two adjustments on them to properly set the meter range so that precise measurements can be taken. It is often necessary to *zero* the meter. This means that when the probes are shorted together, the meter scale will read zero. This adjustment may be found on the meter face itself in the form of a tiny set screw, of the adjustment may be made by a *zero adjust* potentiometer on the multimeter panel. When measuring resistance, it is also necessary to set the scale for the high end. This is done by using the *ohms adjust* control. To properly set this control, it is necessary to first zero the meter with the mode switch in the ohms position and in the appropriate range. With the two probes shorted together, the zero adjust is turned until the meter reads zero. This control is often touchy, and some very fine adjusting may be necessary. Now, separate the test probes and the needle should travel toward the far end of the scale. The last reading on the ohms scale will be infinity, often labelled "inf." This indicates infinite resistance and occurs when the test probes are separated and there is only air in between. By adjusting the ohms adjust control, the indicator is made to read exactly at the infinity marker. Adjusting this control will often have an effect on your meter's zero adjustment, and you may have to reshort the test probes and adjust for a zero reading again. It will then be necessary to go back and touch up the *infinity-adjust* control. Each control has an effect on the other, but after a few short adjustments, the scale should be calibrated so that the needle indicator reads zero when the probes are shorted, and infinity when they are separated. For truly accurate measurements of resistance, this alignment process must be done each time you switch to the ohms position from another point on the mode switch. When moving from one ohms scale to another, this procedure will also have to be repeated.

Never apply the probes of the multimeter to a voltage circuit under power when the mode switch is in the *ohms* position. This can result in damage to the meter. When the mode switch is in the current-measuring position, be extremely careful where you put the probes when testing the circuit. If the probes are placed in parallel with a voltage circuit, the meter will act as a short circuit. This would probably cause severe damage to the multimeter and might even damage delicate components within the circuit under test. For current measurements, an ammeter must be placed in series with the circuit under test. This is true when using the multimeter, which, in the current measuring position, is an ammeter, or to be more accurate, a milliammeter and a microammeter as well as an ammeter in some cases. The black probe from the multimeter is the negative probe and is connected to the positive point in the circuit which has been broken to allow for insertion of the ammeter. The red probe is the positive one and is placed across the other side of the open contact point to complete the circuit for proper current measurements.

As mentioned previously, if you are not certain of the ballpark range of current values within a circuit being tested, set the current scale to the highest position with the value selector switch. When the current is activated, if no reading is shown, adjust the range switch to a lower value until the proper scale is found. The most accurate measurements will be obtained with the multimeter and most other meters when a range is chosen which places the indicator in the upper half of the meter scale. This applies only to conventional, pointer-type meter indicators. Digital readout meters will normally provide accurate readings over their entire range.

To sum up, a multimeter will measure all practical values of resistance, voltage, current, and decibels. This latter measurement is actually a determination of sound level and is the equivalent of a measurement of ac voltage. To get these measurements, however, the test probes have to be connected into the circuit in the correct manner (parallel or in series), and you must know the proper setting in which to place the multimeter mode switch. Most of this knowledge comes from an understanding of the values you wish to measure and their relationship in the circuit.

RESISTANCE MEASUREMENTS WITH THE MULTIMETER

You already know that resistance is measured in ohms and indicates the resistance to the flow of current in an electronic circuit. In order to accurately measure resistance, you must first understand how resistances can combine in a circuit to produce different or unusual values.

Figure 5-3 shows two resistors valued at 100 ohms each in a parallel circuit. When the probes of an ohmmeter are placed across each of these resistors as a single unit and not connected as shown, the meter will indicate 100 ohms. But when resistors are connected in parallel, the total resistance of the parallel circuit will be less than

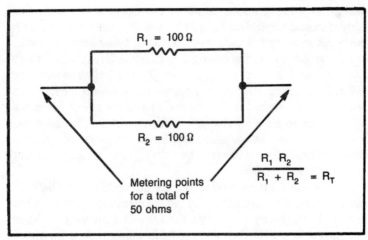

Fig. 5-3. Parallel connection of resistors.

the resistance of any one component. A practical formula for determining resistance in parallel is also shown in Fig. 5-3. Substituting the values for R with the actual values used in the circuit of 100 ohms, the formula works out to 100 × 100, which is equal to 10,000 divided by 100 + 100, or 200, for a total of 50 ohms. The actual value is a fraction of an ohm less than this formula indicates, but is is as accurate as most applications require.

Resistances in parallel offer two paths for electric current to flow. This is the reason that the combination offers less resistance than a single resistor of the value of any one of the resistors combined in parallel. When measuring resistors that are combined in the circuit with other resistors, you must remember that some of the other components will often be wired in a parallel configuration with the resistor under test and will affect the readings you obtain. It may not be possible to measure the true value of a single resistor until it is disconnected from the circuit by clipping or desoldering one of its leads. This effectively removes the resistor from the rest of the electronic circuit, and it may be measured as a discrete component.

Figure 5-4 shows three resistors in parallel. The same formula can be used by substituting the values of R_1 and R_2. The answer from this formula is then substituted back into the formula again for the value of R_1 and the remaining resistor value is substituted for R_2. When the formula is worked again using these values, the total resistance of the parallel circuit of three resistors will be known.

Figure 5-5 shows the proper placement of the multimeter probes to measure a circuit composed of parallel resistances. This is a very

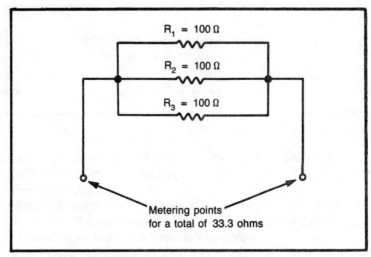

Fig. 5-4. Three resistors in parallel.

easy circuit to measure because the probes are placed across only a single resistor, one probe on one lead and one probe on the other. Although the probes are shown attached to only one resistor, the leads of any other resistor in the parallel circuit may be used. The indicated reading will remain the same.

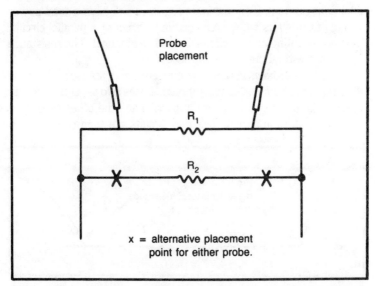

Fig. 5-5. Probe placement points for measuring total resistance in a parallel resistor circuit.

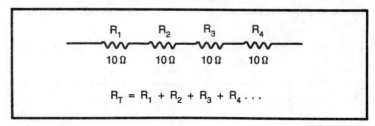

Fig. 5-6. Resistors in series.

When resistors are combined in *series*, the resistance values *add*. Figure 5-6 shows a circuit using four resistors in series. Each resistor has a value of 10 ohms; therefore, the total resistance is 40 ohms. The formula for determining total resistance in a series circuit such as the one shown is: Total resistance = $R_1 + R_2 + R_3$ .... This formula applies for the number of resistors in the series circuit. Figure 5-7 shows the proper connections for measuring the total circuit resistance with a multimeter. The same figure also shows the placement for measuring a single resistor in the series string. You may position the probes in such a way as to measure any number of resistors in a string without removing them from the remainder of the circuit. Although this does not pertain to multimeter measurement, resistors which have stated power values will add those values when combined in the circuit in either a series or parallel configuration. Ten one-watt resistors will have a total power rating of ten watts when combined in a series or a parallel circuit. The power rating of both circuits would be identical. The resistance value of the total circuit would be different.

Figure 5-8 shows resistors combined in a *series-parallel* circuit. This is a more complex configuration that uses two separate circuits; one consisting of two resistors in parallel and the other consisting of two resistors in series. The two circuits are then connected in

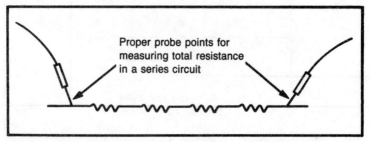

Fig. 5-7. Probe placement points for measuring total resistance in a series circuit.

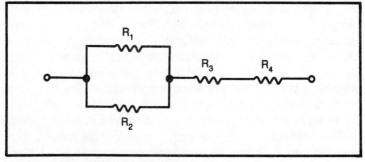

Fig. 5-8. Series/parallel resistor circuit.

a series-parallel combination. It is a very simple task to figure the total resistance of this circuit by treating each of the two circuits as separate entities. Determine the total resistance in each circuit and then determine whether the two circuits have been combined in a series or parallel configuration. Figure 5-9 shows two possible combinations. In A the two circuits are combined in parallel, while in B they are combined in series. Both are series-parallel circuits.

The total resistance of each of these circuits may be measured quickly and accurately through the use of a multimeter. This obviates the necessity for doing the arithmetical calculations. Figure 5-9 also shows the correct placement of the ohmmeter probes for determination of total circuit resistance. The actual measurement will be very close to the total resistance obtained with the formula.

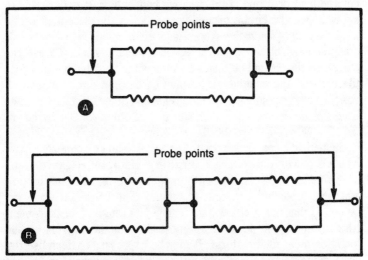

Fig. 5-9. Two different types of series/parallel combinations.

Other components in the circuit may dramatically affect the accurate measuring of resistance. For example, a transformer connected across a resistor will cause the resistance to read almost zero regardless of the value of the component. A solid-state diode or rectifier will cause the resistance to appear very low when the ohmmeter probes are placed across the component leads in a certain way. However, the reading will appear closer to the actual value of the resistor if the ohmmeter probes are reversed. When you want to know the value of a single circuit or component within a more complex circuit, it may be necessary to isolate the circuit under measurement.

In measuring solid-state rectifiers or diodes, an ohmmeter can be very useful in determining if these rectifiers are good or bad. Position the ohmmeter probes across the component leads of the diode. You will get either a very high or a very low reading (usually less than ten ohms but sometimes a bit higher for the low region). In reversing the red and black probes from one lead to the other, you will obtain an opposite reading if the diode is good (very low if the first reading was very high or very high if the first reading was very low). This is just another use of a very versatile instrument.

DC VOLTAGE MEASUREMENTS

The use of the multimeter as an accurate dc voltmeter simply requires the changing of the mode switch to the proper position. The black lead or probe is the negative or ground connection; the colored lead or probe is the positive connection. It is desirable to know where the circuit ground connection is for placement of the negative lead (if a negative ground is used). Many pieces of electronic equipment which are constructed on a metal chassis will use the chassis as the ground tie-point. By placing the negative probe across the chassis, the positive probe may be brought into contact with various voltage points for accurate readings. Some electronic circuits have a *positive ground*. This means that all other voltage points are negative with respect to the chassis. Normally, the chassis will serve as the negative connection point. Figure 5-10 shows a complex circuit which is powered by a nine-volt battery. Here, the negative probe of the meter is placed across the negative terminal of the battery which will usually be the circuit ground. The voltmeter scale will be set so that nine volts will not exceed its range of measurement (the 10- or 15-volt dc scale will most likely be used and is usually found on most multimeters). It may be necessary to decrease the scale range down to five volts or so as various points of the circuit

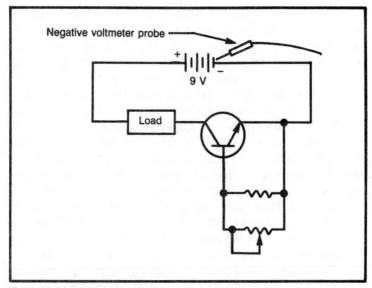

Fig. 5-10. Measuring voltage in a complex circuit.

are probed, because some of the voltages will be dropped in value by the various resistors. But it is reasonable to assume that no potential which is higher than the battery voltage will be encountered if this is the only source of power. Now that the negative probe and the meter range adjustments have been set, it is a simple matter to probe various parts of the circuit for indications of voltage. A dc voltmeter has a very high resistance between the two probes and thus has very little effect on the circuit which it is connected to. Ammeters are just the opposite; they must be connected in series with the circuit and so have very low internal resistances in order to have as little effect as possible on the operation and measurement of the circuit. Experimenters are forewarned that an ammeter connected across the negative and positive terminals of a battery will create a short circuit because the two points are connected by the equivalent of a very low-resistance conductor. A voltmeter has a high internal resistance and is the equivalent of a very large value resistor being connected across the two terminals. Only a tiny amount of current will flow. In a voltmeter circuit, this current is sampled and is converted to a voltage reading on the meter scale. It is easy to see that it is not possible in most circuits to cause circuit damage from taking voltage measurements. This is provided that you don't accidentally short-circuit some components by having the metal tip of the probe accidentally touch two circuits at the same time.

Another note of caution: when measuring medium- and high-voltage circuits, always connect the negative probe to its negative source so that you don't have to hold the probe in your hand. When using the positive probe, use whichever hand you are comfortable with and put the empty hand in your pocket. This could prevent serious injury or death. If your left hand should be resting on the metal chassis (which is circuit ground) and your right hand with the positive probe should accidentally come into contact with a medium- to high-voltage source inside the equipment cabinet, the full potential would pass through your heart by traveling through your arm and chest to complete the circuit. A severe electrical shock can often cause the heart muscle to be temporarily paralyzed and the heart stops beating. If someone is not present who can perform external heart massage, then you, the unfortunate victim, could die. By keeping one hand in your pocket at all times when taking voltage measurements, any electrical shock you may receive will be limited to only one side of your body and cannot travel through the heart muscle. The chances of a serious accident are greatly reduced.

When probing different voltage sources within a piece of electronic equipment, you will note how these values vary. The internal resistance of various components in the circuit is designed to reduce voltage to a level which is usable by other components. In many circuits both ac and dc voltages will be present at the same time in the same conductors or devices. These voltages can sometimes interfere with multimeter readings, especially if the ac voltage has a very high frequency. Most of these meters read the average voltage as opposed to the peak ac value, which will be considerably higher. Standard house current is 110 volts or 120 volts, depending on the area in which you live. This is the average voltage or the *rms* value, which is an abbreviation for *root mean square*. This is what is normally used for performing power calculations, and is the value most often read by ac voltmeters. The ac voltage range may follow the same pattern as the dc voltage range, or it may be completely different, depending upon the meter. This mode of operation reads ac voltages only and will not give accurate dc voltage readings.

The frequency of the ac voltage will determine the accuracy of readings up to the very low electrical frequency range which is still beyond that of human hearing. Most of these multimeters will not provide accurate readings at the higher radio frequencies.

The ac voltage scale is calibrated so that different multipliers may be used just as with the dc and ohms scales. Many of these meters also contain a decibel scale, which is a relative indication of power gain and power loss. When this scale is calibrated to zero decibels for a specific input of audio voltage, decreasing and increasing this audio voltage will cause a subsequent increase and decrease in the decibel reading which measures audio power gain. This is only a relative indication. If an audio power output of one watt produces a reading of zero decibels, then reducing the power to one-half watt will show a drop of three decibels on this scale. The decibel is a unit which is arrived at by certain logarithmic calculations, but you don't need to know how it is derived in order to use it effectively. Figure 5-11 shows the various ways that a multimeter may be connected to ac circuits to read ac voltages.

CURRENT MEASUREMENTS

Most multimeters also offer the advantage of measuring direct current and sometimes even alternating current. In order to do this, it is usually necessary to move one or two of the probes to one of the current slots. But some models perform this changeover on the mode switch. The directions for the particular model you are using will dictate how the initial setup is accomplished. In any event, it

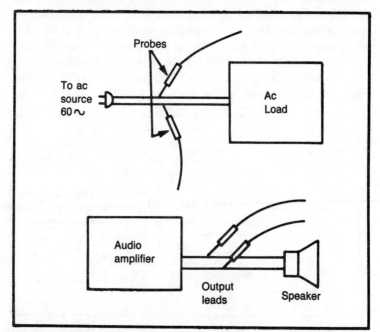

Fig. 5-11. Reading ac voltage with the multimeter.

will be necessary to change the mode switch so that it is in the current position. You will also have to choose the scale which best fits the anticipated range of current you will be measuring. If you exceed the current limitation imposed by the range switch, you may burn out a shunt element in the meter or even destroy the meter itself. Either condition usually necessitates factory repair.

Most multimeters offer a 0 to 100 microampere scale, a 0 to 1 milliampere scale, and a 0 to 1000 milliampere scale. This is a normal range found on multimeters, although some will vary. Most multimeters also contain a high-amperage scale which may read from 0 to 5 or 0 to 10 amperes of current.

It is necessary to know a bit about the circuit under test before attempting to make current measurements. A schematic diagram will almost always be necessary to measure the different current demands at various legs of a complex circuit. It will be necessary to break the circuit at the point where a reading is to be taken in order to insert the ammeter into the circuit in series. The ammeter then becomes a part of the circuit, and its reading increases as the current demand increases and vice versa. Figure 5-12 shows the ideal placement position for an ammeter when a reading of the total current drain of a circuit is desired. Here, the break in the circuit is made at the power-supply input. The positive connection is open and the negative terminal or negative probe of the multimeter is connected to the positive terminal of the battery. The positive meter probe is connected to the circuit where the positive battery contact would normally connect. When the power switch on the equipment is thrown in the ON position, current will flow from the

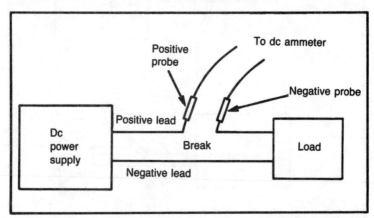

Fig. 5-12. Ideal placement position for an ammeter when reading total current drain.

battery and through the meter. An indication will be given of the amount of current the circuit is drawing in normal operation. The negative terminal of the battery will be separated from its connection to the circuit and the ohmmeter connected at this point with the positive probe. Then the negative terminal of the battery and the negative probe are connected to the remainder of the circuit. Either location is perfectly acceptable and the placement of a meter in a series circuit is not critical. It will give the same measurement of current at any point in the series circuit.

If your meter has an ac amperage scale, it may be connected in exactly the same way as the dc ammeter except that the power source could not be a battery. It would have to be some other source of alternating current. Because an ac ammeter is not a polarized device, the polarity of the meter does not have to be considered.

The main consideration when connecting an ammeter to a circuit is to make certain that the proper polarity connections are observed (where necessary) and that the meter is always installed in series with the circuit that is to be measured. Remember, a parallel connection in a circuit can cause damage to the meter and possibly blow fuses as well.

An ammeter can be an important device in determining how a circuit is operating. Some electronic circuits may appear to be operating properly but may actually be drawing too much current due to a defective component that may eventually cause complete circuit failure. When the normal current drain is known, a multimeter used as an ammeter can be used to detect the high current drain. It can also determine if the current is higher than normal if the proper current drain at various portions of the circuit are known. This applies to complex circuits which may have two or more current paths within the master circuit. The overall current drain in such a circuit may still be read by an ammeter placed in the power-supply input connections as is shown in the last illustration.

In parallel circuits which are all fed from the same power supply, an ammeter may be placed in one of the parallel legs to read the amount of current which is being drawn. Figure 5-13 shows a circuit which consists of three resistors in parallel along with the connection points for the probes from the multimeter to measure current in a particular leg. To measure the overall current of this circuit, the meter would be placed at the power-supply input point. To measure current in another leg of this circuit, a break in the circuit which corresponds to the same break in the leg illustrated would be made in one of the other legs. By knowing your circuit thoroughly

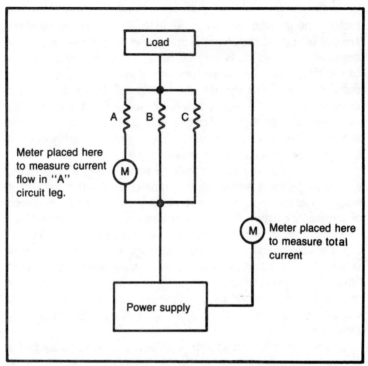

Fig. 5-13. Measuring overall current in a complex circuit.

and by knowing which one you wish to measure and where it is physically located, you can measure all types of current paths in the most complex electronic circuit.

Circuits which contain both alternating and direct current cannot be measured using a dc ammeter unless this leg of the circuit is measured at a point which is not subjected to the alternating current. For instance, in a radio-frequency amplifier circuit, dc current is fed through a *radio-frequency choke*, which is a device that prevents the flow of rf in a circuit. At the plate connection of the amplifier tube, the direct current which is provided by the power supply will be present along with the alternating current which is generated by the amplifier. You cannot insert the ammeter at this point because both ac and dc are flowing in the circuit. You would have to measure the direct current at a point in the circuit between the power supply and the rf choke, because at any point past the rf choke, alternating current at radio frequency is present along with direct current.

144

SUMMARY

It can be seen that the multimeter is an invaluable instrument for a myriad of electronics uses. It is an essential device for any electronics bench and takes the place of many separate instruments. The multimeter allows the home builder to test and align many types of electronic circuits. As a troubleshooting aid, it is really without equal.

The multimeter is usually the first electronic instrument which is obtained by both home experimenters and the electronics technician alike. It is the most-used piece of equipment in the shop, so its many uses should be fully understood. The purchase of a good-quality multimeter is an excellent investment. A poor-quality meter may actually do more harm than good by giving misleading and inaccurate readings. As long as you are involved in the measuring of electronic circuits, you will continue to use the multimeter over and over again.

wide range of

Chapter 6

Electronic Test Instruments

Many electronic measuring devices incorporate sophisticated internal circuits that make measurement and testing a much simpler procedure on many different types of equipment. These devices use transistors, diodes, integrated circuits, and many other electronic components to create a test circuit which has many uses. This chapter will discuss the various test instruments that are likely to be encountered. The oscilloscope, another complex electronic test instrument, is discussed in a later chapter.

TRANSISTOR TESTERS

It is possible to check many solid-state devices by using a multimeter. However, these tests are very limited. The multimeter can tell you if a transistor is good or bad when checking these components, but it can't really tell you the operational properties of the transistor under test. In other words, if you suspect a blown transistor, a multimeter will probably identify this condition. But if a transistor is not operating up to design specifications in relation to frequency response, current gain, etc., the multimeter is almost useless.

For these sorts of tests, a *transistor checker* should be used. These are not especially expensive devices and they offer many test features which cannot be found on even the most expensive multimeter.

There are many parameters, conditions, and readings associated with transistors and related components. The transistor checker is available from many different electronic companies and most will give good indications of not only good-bad component conditions, but will also measure leakage, current gain and even identify the three component leads for you. Some perform only a few tests, while others will perform many simultaneously. Some of the better models do not require the connection of coded instrument leads to specific component leads. In these latter models, any of the three leads emanating from the transistor checker may be connected to any of the three leads of a bipolar transistor.

Figure 6-1 shows a popular transistor checker made by Dynascan Corporation. This is a very simple device to use and may be powered from the ac line or from an external 12-volt power source for use in the field.

When checking a bipolar transistor, all you do is connect the three checker leads (colored yellow, blue, and green) to the three protruding leads of the transistor. With the checker turned on, a

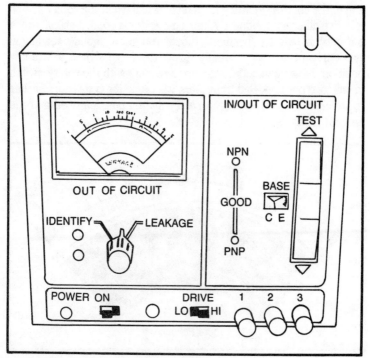

Fig. 6-1. A test bench transistor checker.

large rotary switch is turned to several different positions. If the transistor is good, one of these positions will cause the transistor checker to emit an audible signal. This indicates that the transistor is functioning properly. Note that next to the rotary switch is a small window. As the switch is rotated, the yellow, blue, and green colors are rotated in the window. Each of the color positions is marked with "collector," "emitter," and "base," respectively. This window is shown in Fig. 6-2. If the color yellow is in the collector position, this means that the yellow lead from the transistor checker is connected to the collector of the transistor under test. The same matching is done with the blue and green colorations to determine the emitter and collector leads of the transistor.

Transistor leakage is tested by rotating an automatic return switch beneath the meter window. This window and switch are shown in Fig. 6-3. A transistor can check out as good in this circuit and still malfunction in another circuit if leakage is high. A drive control is also supplied on the Dynascan transistor checker. Most of the time (especially for large-signal transistors), transistors are tested in the high-drive position. Sometimes, however, this will result in the checker emitting its audible signal when the rotary switch is in two different positions. When a transistor is good, it should check out good in only one position. Should you get a "good" reading in two different positions, simply throw the drive switch to the low position. Now you should get a good reading with the rotary switch in only one position. Bad transistors will cause the transistor checker to be silent regardless of the rotary switch position used and the amount of drive.

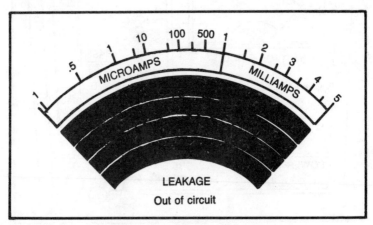

Fig. 6-2. A meter is provided for measuring leakage current.

148

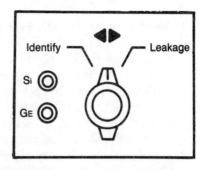

Fig. 6-3. A switch is provided to test for leakage in one position and to identify polarities in the other.

I have used this type of transistor checker in many applications and find it to be an excellent unit for all testing purposes. It can even be used to test diodes and some silicon-controlled rectifiers (SCRs).

Most manufacturers of transistor checkers indicate that their instruments will test transistors while they are wired into their respective circuits. In other words, it is not necessary to remove the transistor for testing. This statement is true for most situations, and these units will test transistors in circuit, but they will not test *all* transistors in a circuit. I have found that it is usually necessary to remove the transistor from the circuit to get an accurate reading. Most of the time, all that is necessary is to remove two of the transistor leads from the circuit that contains the component under test. With any two leads isolated from the circuit, the entire component is isolated and may be tested by clipping the checker leads to the transistor leads as explained earlier. Good-bad testing can often be accomplished in-circuit with most of these checkers, but it's impossible to get an accurate leakage reading.

For those who intend to check transistors in-circuit, caution is essential. Make absolutely certain that no operating voltage is applied to the transistor while under test. If voltage is present on the transistor while it is being tested, your checker may be damaged. This has happened to me several times while testing radio transmitters. These tests often require that the transmitter be activated for a short period and then turned off before applying the checker leads. Several times the power deactivation step was forgotten and each time a diode was destroyed in the transistor checker. Should the checker leads inadvertently be connected to a moderate voltage source, far more serious damage can result.

One drawback of the transistor checker just discussed is its size. Although a very compact instrument which can easily be held in one hand, it is too big to fit in your coat pocket for field testing. Also, if you are away from a 115-volt ac source, it is necessary to strap on an external battery pack. Field testing of electronic components may require many different instruments and those which are microminiature in size are especially practical. For field-testing purposes, I use the Hickock transistor/diode checker shown in Fig. 6-4. This instrument does not offer the full range of tests offered by the previously discussed model, but is handy in that it can always be carried with you. It will easily fit into a shirt pocket.

Figure 6-5 shows the instrument in the process of testing a bipolar transistor which was pulled from a piece of electronic equipment. The transistor is simply inserted into a socket provided on the chassis of the tester. When the two-position switch is thrown to the test position, three panel lights begin to blink in rotation if the component is defective. If the component is good, the lights will cycle once and only one will remain lit. These lights are actually light-emitting diodes and use very little current. The light which remains on indicates the base lead of the component under test. This leaves the collector and emitter of the transistor, which must be guessed

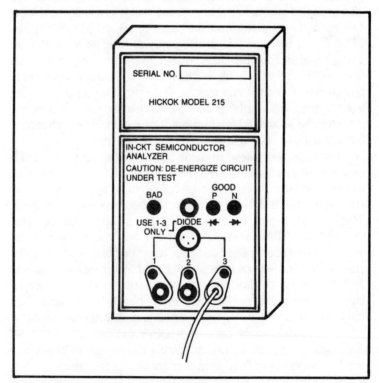

Fig. 6-4. The hand-held Hickock transistor checker.

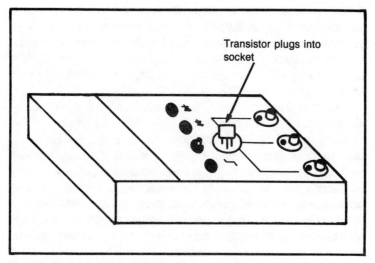

Fig. 6-5. Testing a bipolar transistor.

at by the technician. This is not as difficult as it sounds, because transistor emitters are usually identified on the component case. Most bipolar transistors of the metal can variety use the component case as the collector connection. A simple check with an ohmmeter with its probes connected between the transistor case and switched between the two remaining leads will quickly indicate which lead is connected to the case.

The Hickock meter is powered by two internal nine-volt batteries of the transistor-radio variety. These will provide many hours of intermittent service which is typical of transistor checker use. In many applications, the batteries will provide shelf-life operation.

Transistor checkers are in wide use for most types of electronic testing and measuring today. As noted previously, these instruments are relatively inexpensive and cost anywhere from $20 to about $250 for standard units. Even the least expensive models are usually far superior to a multimeter for testing the performance of solid-state devices. The ohmmeter is still used occasionally for transistor testing but only in emergency situations where a transistor checker is not available.

TUBE TESTERS

While the use of electron tubes in modern equipment is rare these days, the technician will often encounter many older circuits which still use these devices. The vacuum tube is a very complicated

151

device compared to the transistor and is subject to many problems. Because elements are suspended within the glass envelope, these can open up or become shorted. The filament can open or wear out after many hours of use. The envelope can leak and cause the tube to become gassy. This will result in improper operation or no operation at all.

At some time in his career, every technician has needed to check an electron tube. The ohmmeter can be called to service for this task and will indicate both open and shorted elements. However, the ohmmeter cannot detect a gassy condition or a filament which no longer emits electrons. There are many types of tube testers on today's market. Some of them may be advertised as plate conductance, transconductance comparison, mutual conductance, or dynamic output types. Regardless of how their circuits operate, most tube testers can be put into one of two categories, either *emission testers* or *transconductance testers.*

Emission-type tube testers indicate the emission capabilities of the cathode in a tube under test. To check for emission, the tube is usually connected in a diode rectifier configuration by the tube-tester circuitry. Figure 6-6 shows a typical electron tube. It consists of three grids, a plate, and filament/cathode. When connected to the tube tester, the internal circuitry of the tester effectively shorts the grids and the plate. With the proper filament voltage and low plate voltage applied, a milliammeter in the plate circuit reads the cathode current flow. All tube testers are equipped with manuals which indicate the proper current flow for each type of tube or contains specific setup information when the tester is being used with a certain tube. When the latter is used, a good/bad designation appears on the meter face. If the needle falls in the good position, the tube checks out all right; if it is in the bad area, the tube should be replaced.

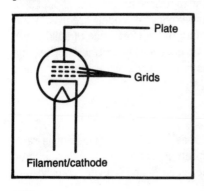

Fig. 6-6. A pentode electron tube.

152

The emission method of tube testing gives a fairly accurate indication of a tube's condition. A fall-off in emission is usually a sign that the tube has reached the end of its useful life. Sometimes, the cathodes of electron tubes will develop hot spots which emit far more electrons from a small area than is normal. This emission is so great that the section of grid which lies in parallel with these spots is unable to control the electron stream. In these circumstances, the total emission of the cathode will be normal, but the tube will not operate properly in a circuit. Here, the emission type of tube tester will probably give a "good" indication, although the tube is actually bad.

Most tube checkers sold today are the conductance type. These are able to provide more reliable checks of electron tubes and theoretically tell you more about the overall tube condition.

Instead of shorting tube elements together, the conductance type of tube tester allows the tube to operate as a simple amplifier. This is the usual purpose of most electron tubes and their output can determine just how well they are operating. Transconductance tube testers supply appropriate plate, screen, and bias voltages in addition to filament voltage. When properly set up, the tube tester with the electron tube forms an amplifier. A milliammeter is connected to the amplifier output to measure gain.

Figure 6-7 shows a simplified schematic drawing of a typical transconductance tube tester. It is necessary to be able to vary the

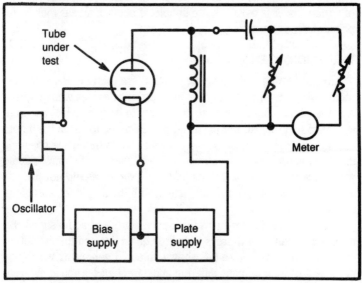

Fig. 6-7. Basic circuitry of a transconductance tube tester.

supply voltages provided by the tube tester in order to be able to check many different types of electron tubes. A tube manual is included with these testers to indicate the proper control positions for different types of tubes. Often, a rolling chart is built into the tester chassis. A tube tester is next to useless if you lose the setup manual, so a built-in chart is a good idea.

Regardless of whether it is the emission or the conductance variety, all tube testers have advantages and disadvantages. I indicated earlier that the transconductance tube testers provided more accurate information about the tube under test than the emission type, but this is not always the case. Some high output amplifier tubes may check out as normal under the low current drain which is used in the tester. But when the tubes are placed back into their amplifier circuits where high amounts of current are being drawn, they may not operate due to low emission. Here, an emission checker might be the better instrument. Then too, transconductance checkers are far more expensive than the emission type due to their more complex internal circuits. The emission type needs only to supply operating voltage and has few other internal components. Since transistors have all but replaced electron tubes, there are many tube checkers available (some for a few dollars) through electronic surplus channels. The careful shopper will often find excellent buys in tube checkers. Some of these cost hundreds of dollars when new and are usually in excellent working condition. Since the internal circuitry of a tube tester is usually quite simple, it is often an easy job to repair a defective unit.

THE AUDIO GENERATOR

When testing audio amplifiers which are found in PA systems, radios, stereos, and other types of electronic devices whose output is in the audio range, it is often necessary to have a stable, accurate source of audio signal. For example, if you wanted to check the output of a PA system, it would be very difficult to do this while speaking into the microphone. The human voice is a series of complex audio waves that vary in magnitude. An output meter would not provide a continuous reading, but would travel with the amplitude variations of the human voice.

An *audio generator* provides a pure source of audio signal. By pure, it is meant that there is very little distortion and that its output remains constant. Figure 6-8 shows an audio generator which will produce an output at many different audio frequencies and at different levels. This particular model is typical and can provide accurate

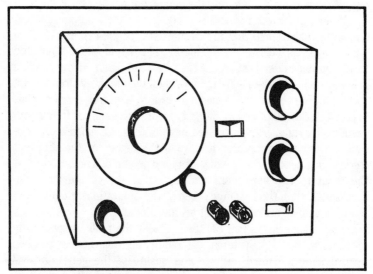

Fig. 6-8. An audio generator.

outputs at frequencies ranging from 30 Hz to over 10,000 Hz, which is beyond the hearing range for most adults.

There are various different controls on this audio generator. One sets the frequency range (say, from 1000 to 5000 Hz). Another is a fine adjustment which will vary the frequency in ten-hertz steps. Still another is used to control the output level of the signal being generated. This output level is often indicated on a meter on the audio-generator panel.

The audio generator is often called a *tone generator*. This is a fairly accurate description in that these devices do generate audible tones. To hear them, all that is necessary is to connect a small speaker to the generator output. Electronic organs and other similar types of musical instruments are merely very complex audio generators.

Now, when testing the same PA system, the output of the audio generator is connected to the microphone input. A steady tone of 1000 Hz (for instance) might be fed to the PA circuit. The output can now be accurately read because the input is in a steady state.

It will be necessary to adjust the output of the audio generator when different frequencies are selected. If an in-circuit meter (rated in decibels) is provided in the generator, this is used to accurately set the output at different frequencies. When testing the frequency response of a circuit, you might start with a 1000-Hz tone at an indicated output of −10 dBm, as indicated by the meter. When an

155

output reading is obtained from the equipment being tested, the tone would then be lowered to 100 Hz. The generator's output controls should be adjusted so that the circuit delivers the same − 10-dBm output, again as indicated by the meter. The output meter connected to the circuit under test would then be read again. If the reading is lower than with the 1000-Hz tone, then the amplifier circuit is not as responsive to low frequencies as it is to high frequencies. This is just an example of how audio generators are used for frequency-response testing. The main idea here is to make certain that the output from the generator is always the same for the different frequencies used during testing. As long as you can be certain that the input to the equipment under test is always the same (in amplitude), then the readings taken at the output meter connected to the amplifier under test will be meaningful.

Audio generators can be used to test the audio amplifiers in stereo systems, PA systems, car radios, and many other devices by providing a known output. The quality of the generator will determine the accuracy of the readings which are finally arrived at.

Some inexpensive audio generators are not equipped with output meters. If you have such a meter, it is a simple matter to measure the output level by using a common ac voltmeter. Figure 6-9 shows how this is done.

The voltmeter is set to read alternating current and is connected across the output terminals of the audio generator. This is exactly what is done in audio generators which have built-in meters, but this arrangement provides an outboard meter. Many multimeters are even calibrated in decibels. It is a simple matter to adjust the generator output for a particular ac-voltage reading at a specific frequency. Using this as a reference, when the generator frequency is changed, adjust its output so that the multimeter reads the same. By doing this you will always be assured that the output from your generator is held constant. Some inexpensive multimeters may respond a bit differently to the high- and low-frequency outputs, but most will have enough frequency response to make this factor negligible.

RF GENERATOR

When testing receiver circuits, it is often necessary to supply a source of radio-frequency energy at the antenna input terminals in order to effect proper alignment. The most convenient method of rf test generation is through the use of an *rf oscillator*. This is

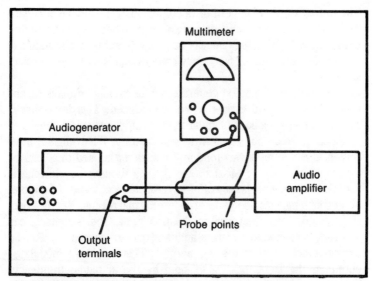

Fig. 6-9. Setup for measuring audio output with a multimeter.

a circuit whose output falls in the radio-frequency range which lies above approximately 100 kHz.

Figure 6-10 shows a typical rf generator which is capable of producing a radio frequency output at ranges from the 100 kHz to UHF band. The UHF spectrum involves frequencies above 300 MHz.

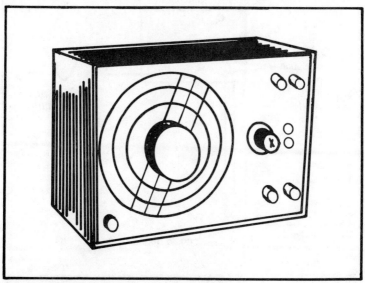

Fig. 6-10. Typical rf generator.

157

The output of the rf generator is continuously adjustable in regard to frequency. Frequency settings are determined by a scale on the front panel. This scale is usually not very accurate in continuously adjustable models and a frequency counter will be needed for exact settings.

The output of most rf generators is *unmodulated*. This means that only a standard carrier wave is produced. This is all that is needed for many jobs, but most rf generators provide a means of modulating the output signal with an audio tone. When aligning a receiver, the volume of this tone in the speaker is used to adjust for maximum sensitivity. Most of the time, a thousand-cycle tone is used, but some models may offer several different tones. Of course, other tones may also be generated by external means. Nearly every rf generator has an external modulation input. These contacts can be used to connect a separate audio-frequency generator, like the ones just discussed, to the rf generator. This provides a modulated output at the frequency established by the af (audio frequency) generator. This setup is shown in block diagram form in Fig. 6-11.

For example, assume that the rf generator is set for an output of 30 MHz and the audio-frequency generator is set at 1500 Hz. This will result in an output from the rf generator which consists of a

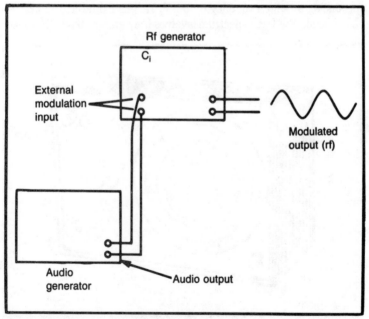

Fig. 6-11. Using an audio generator to modulate the output of an rf generator.

158

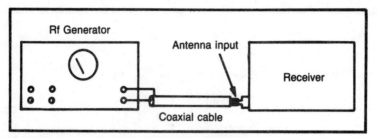

Fig. 6-12. Receiver alignment using an rf generator.

30-MHz signal modulated with a 1500-Hz tone. You can even connect the output of a PA system to the external modulation input and broadcast your voice at any frequency which can be produced by the rf generator.

Figure 6-12 shows a typical setup for receiver alignment using the rf generator. The generator's output is connected to the antenna terminals of the receiver. The latter device is tuned to 30 MHz (for example) and so is the rf generator. The receiver is then adjusted for maximum sensitivity, as will be indicated on its S-meter or by the strength of the tone in the receiver's speaker. Most rf generators will produce outputs in the television frequency spectrum and may be used for audio alignment here as well.

Some specialized rf signal generators are available today which are crystal controlled. This means that their output frequencies are accurate to very high tolerance levels. Figure 6-13 shows the block diagram of the B & K CB Signal Generator, which is specifically designed for citizens'-band service work. It is intended for testing, troubleshooting, and aligning both AM and single-sideband citizens'-band transceivers. It offers all the necessary signal-generation capabilities for complete transceiver testing, as it generates a highly accurate and stable rf output which may be unmodulated, amplitude modulated, or modulated with simulated single sideband. Each citizens'-band channel is selected on the front panel and the rf output and modulation levels can be continuously controlled.

Other specialized signal generators are available for VHF radio servicing, television servicing, and for many other applications. The higher-priced units will offer better frequency stability for more accurate testing and alignment procedures.

GRID-DIP METER

Another type of rf signal generator which is often used in the testing and alignment of rf circuits is the *grid-dip meter*. This

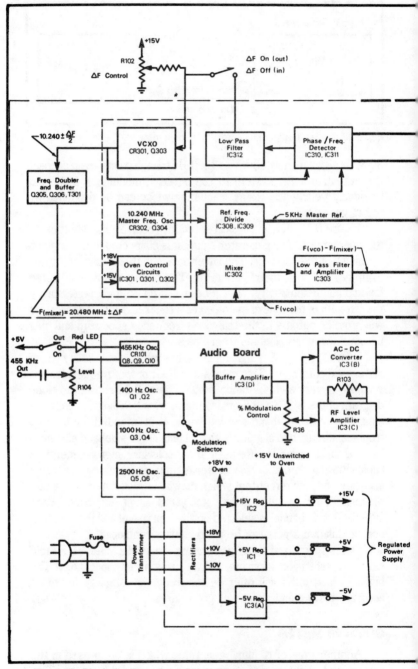

Fig. 6-13. Block diagram of the B & K CB signal generator.

160

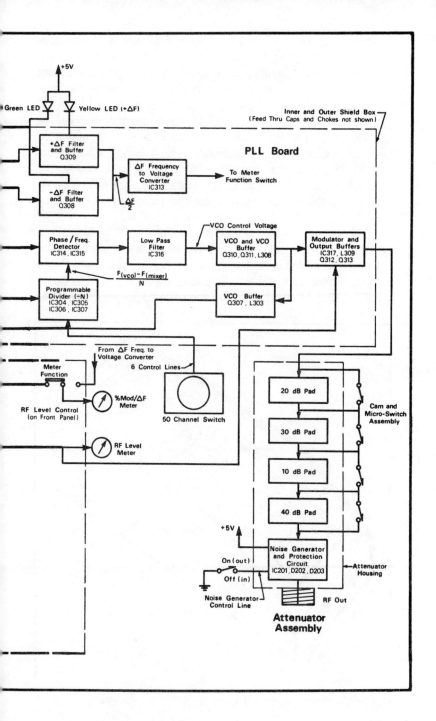

161

instrument, shown in Fig. 6-14, combines the functions of a variable-frequency oscillator and an absorption wavemeter. The grid-dip meter can be used to quickly determine the approximate resonant frequency of any LC circuit which is held in the field of the oscillator coil. Figure 6-15 shows how a grid-dip meter is coupled to an LC circuit. Once the coupling is complete, the frequency control of the grid-dip meter is rotated while reading the self-contained meter. At resonant frequency, the meter indicator will go through a pronounced dip. The frequency at which the dip occurs is then read on the calibrated dial. Most grid-dip meters come with many different coils which allow the instrument to be used at a multitude of frequency ranges.

The grid-dip meter, or *grid-dip oscillator,* as it is sometimes called, is not extremely accurate. The frequency indicated on the calibrated scale should be thought of as only a ballpark figure. More accurate indications of the exact frequency of dip can be had by tuning a calibrated radio receiver to the approximate frequency range as

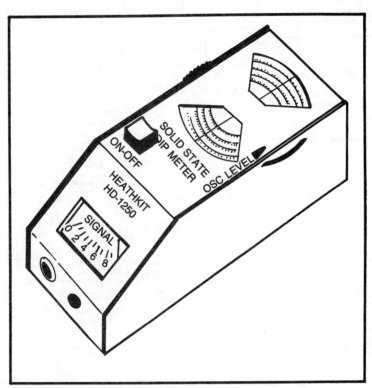

Fig. 6-14. A typical grid-dip meter.

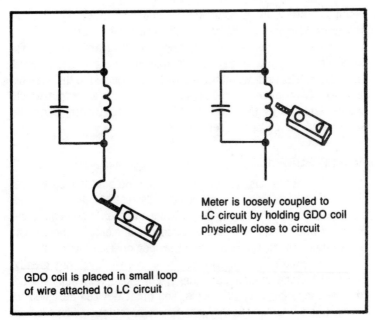

Meter is loosely coupled to
LC circuit by holding GDO coil
physically close to circuit

GDO coil is placed in small loop
of wire attached to LC circuit

Fig. 6-15. Coupling the grid-dip meter to LC circuit.

indicated by the meter's scale. When the receiver crosses the frequency of the grid-dip oscillator, a squeal will be heard. The exact frequency of the oscillator can then be read on the receiver dial.

This latter method for determining the frequency of the grid-dip oscillator is to be preferred for more accurate test purposes. The grid-dip oscillator really cannot be accurately calibrated using its built-in scale due to the frequency instability which is inherent in all of these devices. A calibrated receiver, on the other hand, acts correctly on the grid-dip oscillator signal and is a much more accurate means of determining frequency.

Grid-dip meters can be used to determine capacitance and inductance values. For example, if an LC circuit is made up of a capacitor of known value and an inductance whose value is not known, once the dip frequency is established, the unknown value of inductance can be calculated. The formula for this is:

$$F = 0.189/\sqrt{LC}$$

It will be necessary to use a calculator to solve this equation and you must insert the known value of capacitance. In the above formula, F is the resonant frequency, L is inductance, and C is capacitance.

163

When a known value of inductance is used in an LC circuit, the unknown capacitance can be calculated in the same way.

As a signal generator for alignment of receivers, the grid-dip meter comes in quite handy when a standard rf signal generator is unavailable. The grid-dip meter coil is closely coupled to the antenna terminals of the receiver and adjusted to produce an output which can be detected by the receiver. The receiver circuitry can then be adjusted for maximum sensitivity.

POWER SUPPLIES

While not test instruments specifically, dc power supplies are almost always included at any electronics test bench. Solid-state equipment requires very stable dc voltage in order to function properly, and when troubleshooting electronic equipment, it is necessary to make certain that it is receiving proper operating current.

Power supplies designed for bench service work will vary greatly in output voltage and current ratings. For example, a supply used to power tape players, CB radios, and other devices designed to be operated in a car should have the same output voltage as an automotive battery. Figure 6-16 shows a typical bench power supply which is variable from about 9 V to 16 Vdc. A built-in meter allows for precise adjustments of the voltage level. When testing automotive equipment, the voltage is set at 13.5 Vdc, which is the normal output of the automobile electrical system with the engine running. When testing transistor radios which often operate from nine-volt batteries, the supply voltage is dropped to this value. The meter serves two purposes; first of all, to indicate the output voltage of the power supply; and second, to indicate the current drain of the device being

Fig. 6-16. A typical low-voltage bench power supply.

powered. If current drain is excessively high, this usually indicates a malfunction in the equipment under test. Without this type of power supply and its meter indications, routine service work would be quite difficult.

Other power supplies may deliver up to a thousand volts or more of direct current and are designed for powering specific classes of equipment. Most bench power supplies are categorized into three groups. Low-voltage supplies generally produce an output of from 0 to about 50 volts dc. Medium voltage supplies produce output of from 50 to 350 volts dc. High-voltage supplies produce outputs from 350 to over 1000 volts dc. When testing some types of rf amplifiers, it may be necessary to have a bench supply which will produce 3000 volts or more. This is rare, however, in that most equipment of this type will contain its own built-in dc power supply which operates from the ac line.

OTHER TEST INSTRUMENTS

There are many other test instruments which will be encountered in electronics servicing. Most of these are designed for specialized applications and will differ greatly from manufacturer to manufacturer. Integrated-circuit testers are used to check the quality of various types of integrated-circuit devices. These are relatively simple test instruments but involve a great many switching applications. Specialized field-effect transistor testers will also be encountered. Most of these will also check silicon-controlled rectifiers, triacs, unijunction transistors, and the usual bipolar transistors and diodes.

Capacitor and resistor substitution boxes are also popular. These are simply passive circuits which contain a multiposition rotary switch that will allow many different values of capacitance or resistance to be connected to the output leads. These devices can be thought of as continuously variable capacitors or resistors which can be used for insertion into electronic circuits. Once the correct value of capacitance or resistance is determined, a standard component (capacitor or resistor) of the same approximate value is then chosen for permanent replacement.

For color television work, there are specialized signal generators which produce color bars and dot patterns. These patterns are projected onto the television screen and alignment is effected by visually noting the pattern changes. These generators are often offered as combination units which will include many different test circuits in one housing.

I discussed the multimeter in an earlier chapter. This instrument will take on many different variations from manufacturer to manufacturer. Electronic digital multimeters are quite popular today. Figure 6-17 shows a typical unit. Basically, it operates in the same manner as the earlier models discussed but offers the advantage of value readout in digital form rather than on an analog scale. Digital readout is also available on certain types of wattmeters, SWR meters, and other types of test equipment which are more commonly seen with standard analog meter readouts. Digital readout devices may often be used to test equipment more quickly due to the simplicity of a direct digital indication.

SUMMARY

There are many, many electronic test instruments available to today's technician. While some may appear to be very exotic and complex, each test equipment category operates in basically the same manner. The average technician will usually have need of a multimeter, audio frequency and rf generators, an oscilloscope, a transistor checker, a bench power supply, and possibly a wattmeter.

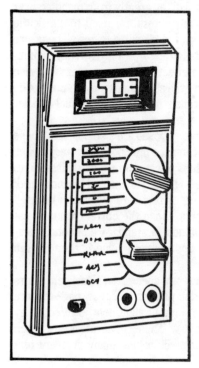

Fig. 6-17. Electronic digital multimeter.

For television service, the specialized generators mentioned previously would also be included in this list of equipment.

When purchasing electronic test equipment, it is certainly not necessary to buy the most exotic and expensive types available. I have a very fine digital multimeter but most of the time I use my trusty, standard Simpson multimeter which is over thirty years old. What you should be looking for, especially when gearing up for the first time, are good, basic instruments of proven design. Most of these will have changed very little in the last two decades. Beware of instruments such as multimeters which look very similar to high-priced models but sell for a fraction of their normal cost. Inexpensive instruments often do not have the same accuracy as those which are moderately priced. In many applications, a reading error of 10 or 20 percent will make little difference, but in others this can be disastrous. It is always important to know the rated accuracy of any piece of test equipment so that this may be taken into account when making critical tests and measurements.

Most equipment manufacturers will be happy to send a complete information packet on any of their products. By studying this and comparing it with ratings of other manufacturers, it is quite simple to make the best choice of which piece of equipment is best suited for your individual needs.

Chapter 7
The Oscilloscope

The oscilloscope is designed for use as a test instrument. It is capable of visually displaying the results of any number of tests on an electronically excited screen. The oscilloscope can be used for a myriad of practical applications. These include alignment of radio and radar receiving and transmitting equipment, hum measurement, frequency comparison, waveform observation, percentage of modulation, and many other similar applications. A block diagram of a typical oscilloscope is shown in Fig. 7-1.

In order to understand the many functions of an oscilloscope, it is necessary to break it down component-by-component. The principal components of a basic oscilloscope include a cathode-ray tube (CRT); a sweep generator; deflection amplifiers (horizontal and vertical); power supply; and suitable controls switches and input connectors. Each of these components will be discussed individually.

CATHODE-RAY TUBE

The cathode-ray tube is the heart of the oscilloscope. As shown in Fig. 7-2, the CRT contains an electron gun, a deflection system, and a screen inside an evacuated glass envelope. It also consists of an intensity control and a focus control. The CRT is the visual display device of the oscilloscope. It operates in the following manner. A beam of electrons leaves the cathode of the tube, is accelerated through the tube and strikes a phosphor-coated screen which glows

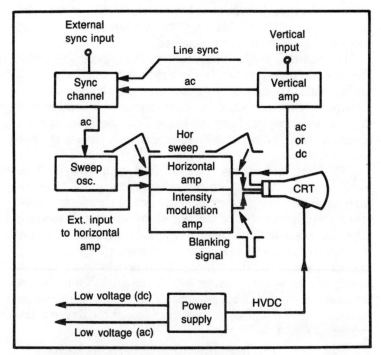

Fig. 7-1. Block diagram of a typical oscilloscope.

at the point of impact. This beam is focused to a sharp point by the action of the focus control. The brightness (or intensity) of the beam is controlled by the intensity control. The beam sweeps the screen in the horizontal and vertical directions when the appropriate voltages

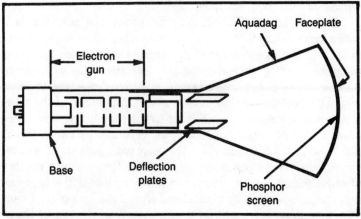

Fig. 7-2. The cathode-ray tube is the heart of every oscilloscope.

are applied to the horizontal and vertical deflection plates. The resultant trace left by the beam is a visual presentation of a voltage waveform in the circuit being examined.

The cathode-ray tube is a special type of vacuum tube in which a beam of electrons is made to strike a phosphor-coated screen and produce light. By moving the beam over the phosphor screen, patterns of light are produced in order to present a visual picture of the voltages used to move the beam. Thus, the electron beam is used to trace pictures of the voltage waveform present at some point in a circuit.

When the cathode-ray tube is used to analyze voltage waveforms, the tube and its supporting circuitry are called an *oscilloscope*. As a test instrument, the oscilloscope is a device whose versatility is limited only by the skill of the operator. Whereas a voltmeter is capable of little more than amplitude measurement, the oscilloscope can be used to measure current and voltage amplitude, frequency, phase and time, in addition to permitting the user to actually see the waveform under test. Distortion in a waveform is generally not apparent from a voltmeter measurement but can be quickly recognized from the pattern traced out on the screen of the cathode ray tube.

Since the electron beam is nearly weightless, the beam can be deflected almost instantaneously, permitting the observation of high frequencies or pulse waveforms of nanosecond duration. The ability of the cathode ray tube to measure minute intervals of time makes it an ideal device for the display and measurement of many types of information. With the addition of intensity modulation, the cathode-ray tube becomes capable of displaying moving pictures and the miracle of television was developed.

The envelope of a cathode-ray tube consists of a large glass bell with a long cylindrical glass tube which is called the *neck*. As shown in Fig. 7-2, the phosphor screen is deposited on the inside of a glass faceplate covering the end of the bell. The electrodes which form the electron beam comprise an assembly which is called the *electron gun*. The electron-gun assembly is placed in the neck of the tube and connects through internal leads to the pins on the base. The beam is moved either horizontally or vertically by two pairs of deflection plates between which the beam must pass on its way to the screen. These plates are mounted on (but not part of) the electron gun. In some types of cathode-ray tubes, the beam is deflected magnetically by a coil placed around the neck of the tube. The magnetically deflected CRT does not use deflection plates. Thus, the cathode-ray tubes are classified as either *electrostatic* or *electromagnetic,* depending upon the method used to deflect the beam.

The inside of the bell of a cathode-ray tube is covered with a conductive graphite coating. This coating, which is called *aquadag,* provides shielding from stray fields which might interfere with the electron beam, prevents light from striking the back of the screen, and most importantly, gathers the secondary electrons emitted when the phosphor is bombarded by the electron stream, returning them to the cathode through the accelerating-anode power supply.

Once the electron beam is formed by the gun assembly, it must travel from several inches to one foot or more before reaching the screen. Because even a small number of collisions between the electron beam and air molecules would adversely affect the operation of the cathode-ray tube, the tube must be highly evacuated. The high degree of evacuation and large surface area of the tube makes the tube especially vulnerable to dangerous implosions of tremendous force. In many cases, sudden jarring or slight nicks or scratches in the glass are sufficient to cause an implosion. Great care should be exercised when handling cathode-ray tubes and their installation or removal should not be attempted without the protection of safety goggles and heavy gloves. When servicing equipment containing cathode-ray tubes, care should be taken that the tube is not bumped or scratched by tools. To further explain the function of the cathode-ray tube and its operation, each part will be described individually.

The Phosphor Screen

The function of the phosphor screen in the cathode-ray tube is the conversion of the kinetic energy of the beam electrons into light or radiant energy. As one of the beam electrons strikes a phosphor atom, its energy is transferred to one of the planetary electrons of the phosphor atom. This causes the electron in the phosphor atom to jump to a higher energy orbit farther from the nucleus. As the atom is thus placed in an unnatural or excited state, the electron will try to give up the excess energy and return to its normal orbit. In returning, the electron releases its excess energy in the form of radiant energy.

The wavelengths of the energy emitted by the phosphor atoms extend from the ultraviolet through the visible spectrum to the infrared. Thus, both light and heat energy are given off by the screen. Should the energy of the beam electrons be too great, the heat developed within the phosphor can alter its chemical characteristics, creating a burned spot on the screen. Once burned, the screen is permanently damaged and will no longer produce light over the area of the burn. When using a CRT, the intensity of the beam should never be greater than that required to produce a usable amount of light on the screen.

The characteristics of the phosphor will depend upon its chemical composition. Phosphors are composed of compounds of zinc, magnesium, and cadmium to which is added small traces of certain other chemicals which are called *activators.* An activator increases the light output of the basic phosphor compounds.

Phosphor screens are classified according to the color of the light produced and the length of time during which the light is given off. The length of time required for the light to decay to one percent of its maximum value after the excitation has been removed is called the *persistency* of the phosphor.

If the persistency of the screen material is about 0.01 second or less, the tube is said to have a short persistency. Tubes with persistency values longer than one second are designated as long-persistency types. If the duration of the glow is between these two values, the persistency is medium.

The emission of light at a temperature below that of incandescent bodies is called *luminescence* (cold light). If light is given off during the excitation of the phosphor, the process is called *fluorescence.* The emission of light which occurs after excitation has ceased (afterglow) is called *phosphorescence.*

Figure 7-3 provides some of the more popular screen coatings and their characteristics. Notice that the various types of coatings are cataloged by letternumber designations such as P1, P4, etc. For example, a P1 phosphor has a medium persistency of 0.03 seconds and has green fluorescence and phosphorescence. In some types of coatings, the fluorescence is a different color than the phosphorescence.

Coating	Application	Persistence	Decay Time (sec)	Fluorescence	Phosphorescence
P1	General Oscilloscope	Medium	0.03	Green	Green
P4	Television	Medium	0.06	White	White
P5	High-Speed Photography	Very short	35×10^{-6}	Blue	Blue
P7	Radar, sonar	Long	3	Blue-White	Greenish-Yellow

Fig. 7-3. Popular screen coatings commonly used in cathode-ray tubes.

The Electron-Gun Assembly

Depending upon the application for which the tube was designed, the electron gun consists of from four to six cylindrically shaped electrodes placed end-to-end within the neck of the tube. For purposes of this discussion, the gun assembly shown in Fig. 7-4 will be used. This tube was designed for general oscillographic applications and has a tetrode-type gun structure. The tube number (5UP1) indicates that the tube has a five-inch diameter screen and a P1 phosphor.

The arrangement of electrodes in the gun and the tube-base connections can be seen in the figure. Because the gun assembly must form the electrons into a pencil-like beam, the electrodes are fabricated in the shape of cylinders. To aid in confining the electron stream to a thin beam, several disc-shaped baffle plates or diaphragms containing small holes at their centers are placed within the various cylinders. These baffle plates serve to screen out all but those electrons travelling close to the longitudinal axis of the gun.

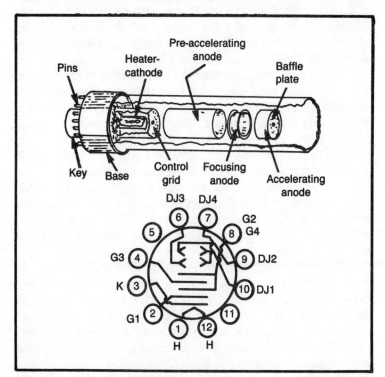

Fig. 7-4. The physical arrangement of components and pin placement of a 5UP1 CRT.

173

The emitter of a cathode-ray tube is indirectly heated and is made in the shape of a cylinder with a flat, closed end. Emission occurs only from the flat end of the nickel cylinder, which is coated with barium and strontium oxides to enhance the emission. The emitting material is maintained at the proper temperature by a tungsten heater placed inside the nickel cathode cylinder. To prevent the twisted heater wire from shorting, the heater is surrounded by a heat-conducting ceramic sleeve. In certain applications, the cathode is directly connected to the filaments to prevent a high-intensity field between the two elements. As shown in the previously referenced Fig. 7-4, the cathode is almost completely enclosed by the control-grid cylinder. This cylinder has a small hole in the center of the flat end through which the electrons can pass on their way to the screen.

When the proper operating potentials are applied to the tube, the pre-accelerating anode is from 1000 to 2000 volts positive with respect to the cathode. This difference of potential causes lines of force to extend from the pre-accelerating anode through the hole in the control grid to the space charge surrounding the end of the cathode. These lines of force cause some of the space-charge electrons to be drawn through the holes in the control grid and accelerated down the length of the tube.

The control grid is operated at a bias potential of from 0 to 90 volts. As the control grid is made increasingly negative, fewer electrons are able to pass through the hole in the grid and the electron density of the beam is decreased. This, in turn, decreases the amount of light produced on the screen. If the control-grid potential is adjusted to approximately −90 volts, electron flow through the grid is entirely stopped and no light is produced on the screen.

By connecting a potentiometer between the control grid and the cathode, the negative grid bias can be adjusted to provide the proper amount of light on the screen. This potentiometer is mounted on the front panel of the oscilloscope and is called the *brightness* or *intensity control*.

The pre-accelerating anode, sometimes referred to as the *pre-accelerating grid,* is followed by the focus and accelerating anodes, respectively. In order to display intricate waveforms without masking fine detail, the beam must be focused to a very small diameter spot by the time it reaches the screen. Focusing of the beam is accomplished by the electrostatic field existing between the focus anode and the pre-accelerating and accelerating anodes.

To illustrate focusing action, the field between the focus anode and the second or accelerating anode is shown in Fig. 7-5. (A similar

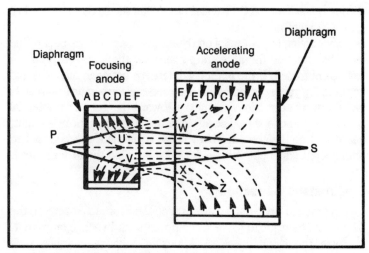

Fig. 7-5. Focus action in a cathode-ray tube.

action occurs between the focus and pre-accelerating anodes). The electrons which succeed in passing through the control grid and pre-accelerating anode come under the influence of the focus anode.

Some of the electrons pass through the hole in the end of the focus anode and into the field between the focus and accelerating anodes. The purpose of the diaphragm is to prevent all electrons except those making a small angle with the axis of the beam from passing through the hole in the diaphragm. This serves to keep the beam narrow. The electrons entering the curved electric field between the anodes are subjected to inward-directed forces, thereby focusing the beam. As the beam passes parallel to the lines of force, the electrons are accelerated to a very high speed. Thus, the net result of the forces influencing the beam of electrons is a high-speed inward-directed beam converging at point S on the screen. The repelling force of like charges in the beam tends to scatter the electrons, but they are accelerated to such a high speed that the scattering action is not effective in defocusing the beam. Nevertheless, the mutual repulsion between electrons in relation to the speed of the electrons determines the sharpness with which a beam may be focused on the screen. The diaphragm on the accelerating anode is used to stop all wide-angle electrons from hitting the screen.

In most cathode-ray tubes, the pre-accelerating and accelerating anodes are connected together internally and therefore operate at the same potential. At an accelerating anode potential of 1000 volts, the focus anode of the CRT under discussion should operate at +170 to +320 volts dc.

By varying the potential on the focus anode with respect to the fixed potential on the accelerating anodes, the focus of the beam can be controlled. If the potential difference between the focus and accelerating anodes is increased, a stronger electrostatic field results and the focal point (point S) moves toward the gun. The potential on the focus anode should be adjusted until the beam forms a bright, sharp spot on the screen. This voltage adjustment is accomplished with a front-panel potentiometer which is called the focus control.

Electrostatic Deflection

In order to trace out a waveform on the fluorescent screen, the beam must be made to move in conformity with a voltage or current. Because the beam consists of moving negative charges, it is surrounded by both a magnetic field and an electric field. If, by some external means, either a magnetic or an electrostatic field is established in the vicinity of the beam, these forces will cause it to shift position. If the movement is to be produced by magnetic means, a coil is placed about the neck of the CRT. As the beam electrons leave the electron gun, they pass through the magnetic field set up by the coil. When the coil field and the magnetic field about the beam electrons interact, the beam is deflected away from its normal position. The amount and direction of beam deflection is determined by the magnitude and direction of the current passed through the deflection coil. Because nearly all cathode-ray tubes designed for use in oscilloscopes use electrostatic deflection, I will discuss this type in some detail.

Electrostatic deflection is accomplished by routing the electron beam between two parallel metal plates to which the deflection voltage is applied. If two parallel plates are positioned near the end of the electron-gun assembly, as shown in Fig. 7-6, the beam can be made to strike the screen at any point along a vertical line passing through the center of the screen.

When no difference of potential exists between the deflection plates, the beam is not affected and will pass directly down the center of the tube and strike the screen at point A. If a difference of potential is applied to the plates, the area between the plates will be filled with lines of force. Assuming the top plate is positive with respect to the bottom plate, an electron in the space between the plates would be attracted by the top plate and repelled by the bottom

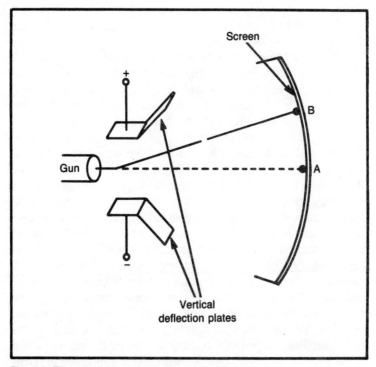

Fig. 7-6. The principle of electrostatic deflection.

plate. Thus, an upward motion would be imparted to the electron by the electrostatic forces between the plates.

Because the electron beam is located between the plates, each electron it contains is acted upon by the force of the electrostatic field. The beam electrons are therefore compelled to follow an upward curving path during the time they are within the influence of the plates. After leaving the vicinity of the plates, the electrons travel in a straight line and strike the screen at point B.

If the amount of voltage between the two deflection plates is increased, the angle of deflection will also increase, causing the beam to strike the screen farther from the center. Any amount of upward deflection can be obtained by applying the right amount of voltage to the plates.

To deflect the beam downward from the center of the screen, the polarity of the voltage applied to the deflection plates must be reversed. The upper plate will then repel the beam and the lower plate will attract it. As before, the amount of deflection is proportional to the magnitude of the voltage applied to the plates.

Horizontal deflection of the beam is accomplished by placing a second pair of deflection plates just beyond the vertical plates so that the beam passes through the two pairs of plates in succession. This is illustrated in Fig. 7-7. The plates that control the horizontal movement of the beam are called the *horizontal deflection plates*. These plates are perpendicular to the plane of deflection. If voltages are applied simultaneously to both pairs of plates, the beam can be moved to any point on the screen.

The deflection sensitivity of an oscilloscope is given as a constant which indicates how much the spot on the screen is deflected (in inches, centimeters, or millimeters) for each volt of potential difference that is applied to the plates. For example, tube specifications may describe a tube as having a deflection sensitivity of 0.2 millimeter per volt dc. This means that every volt of dc applied to the deflection plates causes the spot to move 0.2 millimeter from the undeflected position. Deflection sensitivity is directly proportional to the length of the deflection plates and the distance between the deflection plates and the screen. It is inversely proportional to the separation between the deflection plate and the accelerating voltage.

The *deflection factor* indicates the voltage required on the deflection plates to produce a unit deflection on the screen, and it is the reciprocal of deflection sensitivity. It is expressed in terms of a certain number of dc volts per centimeter (or per inch) of spot movement. For example, for a tube with a deflection sensitivity of 0.2 millimeter per volt dc, the deflection factor is 50 volts per centimeter. It is also common to express the deflection factor in terms of the second anode voltage. For example, 60-volts dc per inch/kilovolt of second anode voltage indicates that with one kilo-

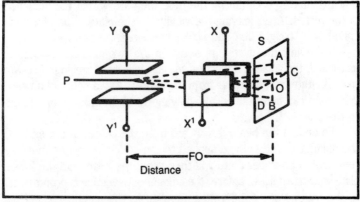

Fig. 7-7. Horizontal deflection of the electron beam.

volt applied to the second anode, the deflection factor is 60-volts per inch. With two kilovolts applied to the second anode, the factor is 120-volts per inch.

The Waveform Display

The eye retains an image for about one-sixteenth of a second. Thus, in a motion picture, the illusion of motion is created by a series of still pictures flashed on the screen so rapidly that the eye cannot follow them as separate pictures. In the cathode-ray tube, the beam is repeatedly swept across the screen and the series of adjacent spots appears as a continuous line. Thus, the waveshape of an ac voltage can be observed on the screen when the ac voltage is applied to one pair of deflection plates and simultaneously a second voltage of appropriate characteristics is applied to the other pair of plates.

The sweep voltage that will produce uniform motion of the spot across the screen is called a *sawtooth wave* because the shape of the voltage waveform resembles the cutting edge of a saw. A sawtooth waveform is shown in Fig. 7-8. The voltage is made to

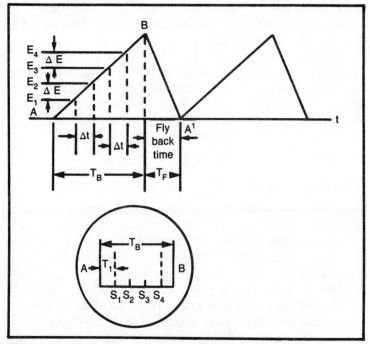

Fig. 7-8. A graphic representation of a sawtooth voltage.

179

rise from point A along a straight line to point B. This is known as a *linear rise.* If this voltage is applied to the horizontal deflecting plates of a cathode-ray tube, the spot will move across the screen to form the time base. The time base will be linear with time if a rise of E volts take place in t seconds anywhere along AB, because that will mean the spot moves from S_1 to S_2 in exactly the same time that it moves from S_3 to S_4. Thus, the sweep is a means of measuring time, since it always takes t_1 seconds to go from A to S_1, or t_4 seconds to go from A to S_4.

It is desirable for the time base to start at the left of the tube, since that is the more usual method of plotting waveforms. The beam is swept from left to right to produce the pattern and must be returned quickly to the starting point to restart the pattern. The beam can be returned quickly only if the voltage falls from B to A' very rapidly. In practice, time T_F is very small compared to the length of the time base T_B. The time, T_F, is called the *fly-back time,* because it represents the time during which the beam is being moved back to the starting point. The fly-back time is so short that the electron beam is swept over the screen too fast to cause emission of much light and the return trace is accordingly very dim. The fly-back time is greatly exaggerated in Fig. 7-8. If the picture were drawn to scale, the time T_F would appear to be almost zero, and the line BA' almost vertical.

If a test voltage from a circuit such as the sine wave in Fig. 7-9 is applied to the vertical deflection plates and the sawtooth sweep voltage is applied to the horizontal deflection plates, the resulting screen pattern will be as shown. As the sawtooth of voltage moves the beam from left to right at a constant rate of speed, the sine wave to be observed deflects the beam vertically. Thus, the sine wave is reproduced on the screen.

POWER SUPPLY

One of the purposes of the power supply is conversion of alternating current (ac) into direct current (dc). In the oscilloscope, the power supply must provide several different values of dc and ac voltages. Direct voltages ranging from 100 to 400 volts must be supplied to the plate and screen circuits of amplifiers. High dc voltages of from -500 to -1500 volts, and in some cases, up to 15 kV, must be provided for proper operation of the cathode-ray tubes. Low ac voltages must be supplied for tube heaters. Figure 7-10 shows a typical block diagram of an oscilloscope power supply.

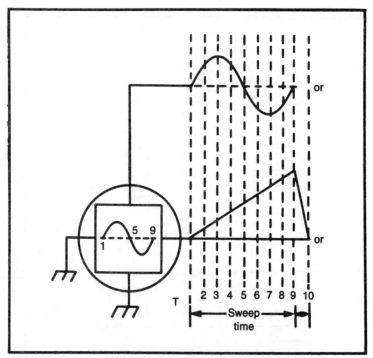

Fig. 7-9. Scope pattern with a sine wave applied to the vertical plate and a sawtooth sweep voltage applied to the horizontal plate.

HORIZONTAL- AND VERTICAL-DEFLECTION AMPLIFIERS

The oscilloscope is a test instrument which must display an exact duplicate waveform of the signal or signals applied to its input. This means that the oscilloscope must not cause any distortion of the applied signals. This is accomplished through the use of linear, class-A amplifiers. The oscilloscope also must have a high input impedance to reduce the loading effect on the circuit or equipment being tested.

The function of both the horizontal and vertical amplifiers in an oscilloscope is to amplify the signal applied to them with minimum distortion. These amplified signals are then applied to the horizontal and vertical deflection plates in the CRT, which will cause horizontal and vertical deflection of the electron beam.

Horizontal Amplifiers

The horizontal amplifier channel is shown in block form in Fig. 7-11. It consists of the horizontal attenuator, horizontal cathode follower, the first and second direct-coupled amplifiers, and the

181

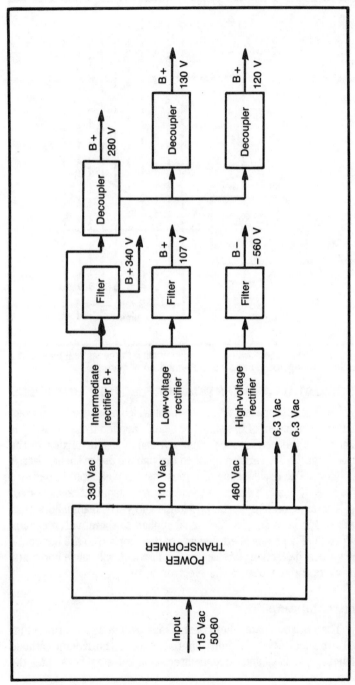

Fig. 7-10. Block diagram of an oscilloscope power supply.

182

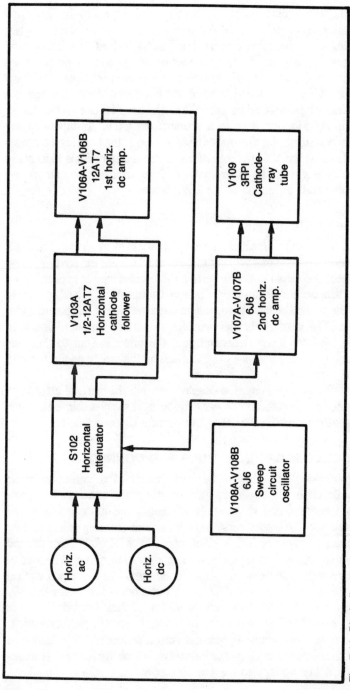

Fig. 7-11. Block diagram of the horizontal amplifier channel.

183

cathode-ray tube. There are three possible inputs to the horizontal channel: two are external and one is internal. In most cases when using an oscilloscope, a linear time base is desired. In this case, the signal fed into the horizontal channel would come from a sweep generator located inside the oscilloscope. In special cases, a signal other than a sawtooth is used for horizontal deflection and the external inputs would be used. The selection of these various inputs is made with the horizontal attenuator switch. All inputs to the channel except the external dc input are fed to the horizontal cathode-follower stage. The purpose of the horizontal amplifier is to increase the strength of the horizontal signal to achieve adequate lateral (horizontal) deflection of the CRT beam.

Vertical Amplifiers

The vertical amplifier channel is shown in block form in Fig. 7-12. There are two possible inputs, both of which are external. In most cases, the signal to be viewed on the oscilloscope is applied to the vertical-deflection amplifier. The specific signal that would be applied to the vertical channel is dependent upon the type of equipment under test. The vertical channel provides two useful outputs. One is to the CRT for beam deflection, and the other is sent to the sync selector for the purpose of synchronizing the sweep generator with the vertical signal.

The ac input signal is coupled through the vertical attenuator to the vertical cathode follower. The dc input signal is coupled through the attenuator but bypasses the vertical cathode follower.

OSCILLOSCOPE CONTROLS AND CONNECTORS

Figure 7-13 shows the front panel of a general-purpose oscilloscope. Oscilloscopes vary greatly in the number of controls and connectors, but generally speaking, the more controls and connectors, the more versatile is the instrument. Regardless of their number, however, all oscilloscopes have similar controls and connectors, and once you learn their operation, you will be able to move with ease from one model to another. Occasionally, identical controls will be labeled differently, but most controls are logically grouped and their names are indicative of their function.

The POWER switch may be a toggle, slide, or rotary switch, and it may be mounted separately or on a shaft that is common to another control, such as the intensity control. Its function is to apply the line voltage to the power supply.

184

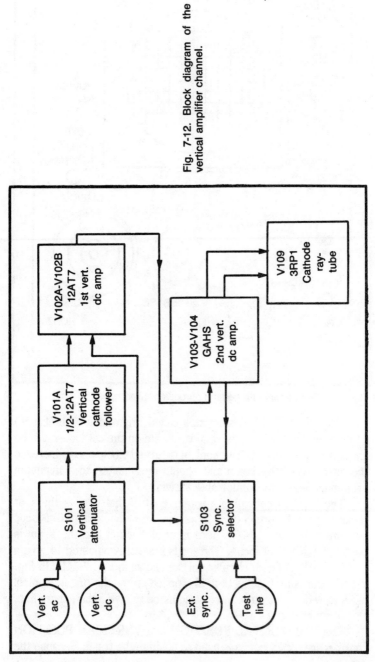

Fig. 7-12. Block diagram of the vertical amplifier channel.

185

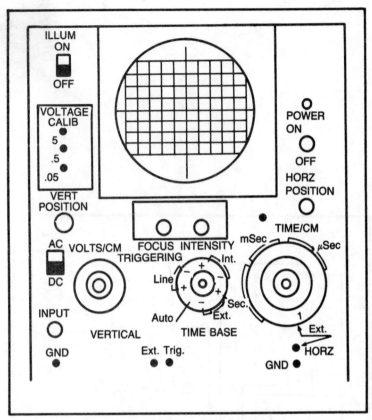

Fig. 7-13. Front panel of a general-purpose oscilloscope.

The INTENSITY (sometimes called brightness or brilliance) control adjusts the brightness of the beam on the cathode-ray tube and is a rotary control. The control is turned clockwise to increase the intensity of the beam and should be adjusted to a minimum brightness level for comfortable viewing.

The FOCUS control is a rotary control that adjusts the spot (beam) size. Figure 7-14 shows the in-focus and out-of-focus extremes. In Fig. 7-14A, there is no deflection and the beam is centered but out of focus. The focus control is adjusted to give a small, clearly defined, circular dot, as shown in Fig. 7-14B. In Fig. 7-14C, horizontal deflection is applied with the focus control misadjusted. The focus control is adjusted to give a thin, sharp line, as shown in Fig. 7-14D.

The HORIZONTAL POSITION and VERTICAL POSITION controls are rotary controls used to position the trace. Because the

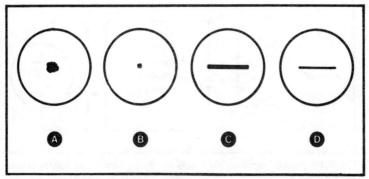

Fig. 7-14. Examples of in focus and out of focus oscilloscope patterns.

graticule is often drawn to represent a graph, some oscilloscopes have the positioning controls labeled to correspond to the x and y axis of the graph. The x axis represents the horizontal movement, while the y axis represents the vertical movement. Figure 7-15 shows the effects of the positioning controls on the trace. In Fig. 7-15A, the horizontal control has been adjusted to move the trace too far to the right, while in Fig. 7-15B, the trace has been moved too far to the left. In Fig. 7-15C, the vertical positioning control has been adjusted to move the trace too close to the top, while in Fig. 7-15D, the trace has been moved too close to the bottom. Figure 7-15E shows the trace properly positioned.

The vertical INPUT (or Y INPUT or SIGNAL INPUT) jack connects the desired signal to be examined to the vertical-deflection

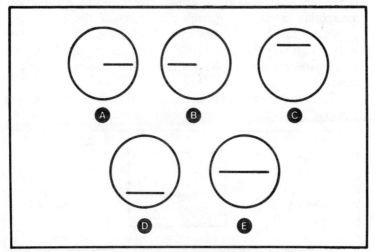

Fig. 7-15. The effects of the positioning controls on the CRT trace.

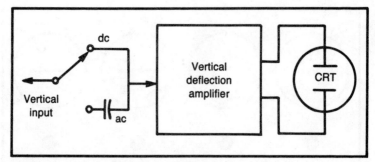

Fig. 7-16. Typical oscilloscope circuit for dc and ac inputs to the vertical amplifier.

amplifiers. On some oscilloscopes, there may be two input jacks, one for ac and the other for dc. Other models may have a single input jack with an associated switch to select the ac or dc connection. In the dc position, the signal is connected directly to the vertical-deflection amplifiers, while in the ac position, the signal is fed through a capacitor first. Figure 7-16 shows the schematic of one arrangement.

The deflection amplifiers increase the amplitude of the input signal level required for the deflection of the CRT beam. These amplifiers must not have any other effect on the signal, such as changing its shape (called *distortion*). Figure 7-17 shows the results of distortion occurring in a deflection amplifier.

An amplifier can handle only a limited range of input amplitudes before it begins to distort the signal. To prevent this, oscilloscopes incorporate circuitry to permit adjustment of the input signal amplitudes to a level that will prevent distortion. This adjustment is the ATTENUATOR control. This control extends the usefulness of the oscilloscope by enabling it to handle a wide range of signal amplitudes.

The attenuator usually consists of two controls. One is a multiposition switch and the other is a potentiometer (Fig. 7-18).

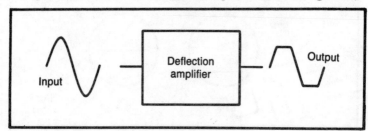

Fig. 7-17. Graphic representation of distortion occurring within an amplifier.

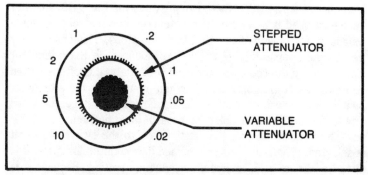

Fig. 7-18. A typical oscilloscope attenuator arrangement.

Each position of the switch may be marked either as to the amount of voltage required to deflect the beam a unit distance such as volts/cm, or as to the amount of attenuation given to the signal, such as 100, 10, or 1.

In the first case, suppose the .05 volt/cm position was selected. In this position, the beam will be deflected vertically one centimeter for every .05 volts of applied signal. If a sine wave occupied four centimeters peak-to-peak, its amplitude would be $4 \times .05 = .2$ volts pk/pk (see Fig. 7-19).

The attenuator switch provides a means of adjusting the input signal level to the amplifiers by steps. These steps are in a definite

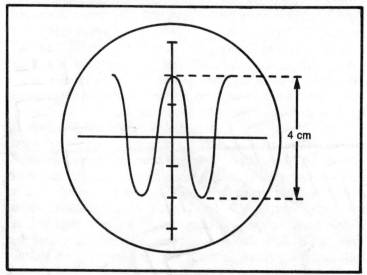

Fig. 7-19. CRT waveform of a four centimeter peak-to-peak input.

sequence, such as 1-2-5, as shown previously in Fig. 7-18. Another sequence used is 1-10-100. The potentiometer control provides a means of fine or variable control between steps. This control may be mounted separately or it may be mounted on the attenuator switch. When the control is mounted separately, it is often marked as FINE GAIN or simply GAIN. When mounted on the attenuator switch, it is usually marked VARIABLE, with the panel markings and knob of a different color for ease of identification.

The variable control adds attenuation to the switch step that is selected. When the control is turned fully counterclockwise, it is at the minimum gain, maximum attenuation setting. Because it is difficult to accurately calibrate a potentiometer, the variable control is either left unmarked, or the front panel is arbitrarily marked off in some convenient units (e.g., 1-10, 1-100). The attenuator switch, however, can be accurately calibrated to the front panel designations. To do this, the variable control is turned fully clockwise to cut it out of the attenuator circuit. This position is usually marked CAL (calibrate) on the panel.

The sweep generator in an oscilloscope serves to develop the sawtooth waveform that is applied to the horizontal deflection plates of the cathode-ray tube. This sawtooth causes the beam to move at a linear rate from the left side of the screen to the right side. This trace, or sweep, is the time base of the oscilloscope. To enable the oscilloscope to accept a wide range of input frequencies, the frequency of the time base is variable. Again, two controls are used. One is a multiposition switch that changes the frequency of the sweep generator in steps, and the second control is a potentiometer that varies the frequency between steps. This is shown in Fig. 7-20. The switch has each step calibrated and the front panel is marked TIME/CM. A 1-2-5 sequence is used for numbering the switch positions and the front panel has markings that group the numbers into microseconds, milliseconds, and seconds.

The potentiometer is labeled VARIABLE and the panel is marked with an arrow indicating the direction to turn the pot to the calibrated (CAL) position. When you want to accurately measure the time of one cycle of an input signal, the variable control is turned to the CAL position and the TIME/CM switch is turned to select the appropriate time base. Suppose the five microsecond position is chosen and two cycles of an input signal are being displayed, as shown in Fig. 7-21. One cycle occupies five centimeters along the horizontal axis. Each centimeter is worth five microseconds in time,

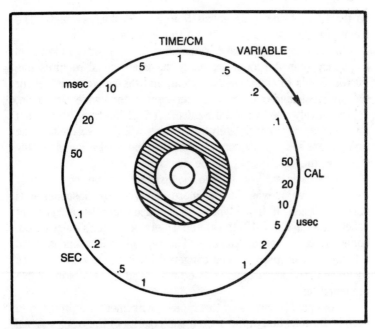

Fig. 7-20. Time-base frequency control.

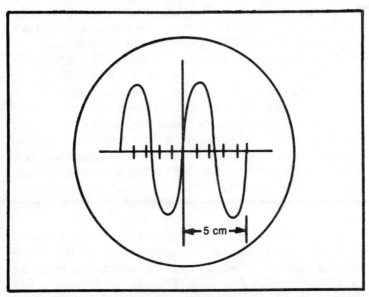

Fig. 7-21. Using the calibrate position and the time-base control to accurately measure an input signal.

so the time of one cycle equals $5 \times 5 = 25$ microseconds. The frequency may be found by using the formula, $f = 1/t$.

In selecting a time base, remember that it should be lower in frequency than the input signal. If the input signal requires five microseconds to complete one cycle, and the sawtooth is set for 0.5 microsecond/cm with a ten-centimeter wide graticule, approximately one cycle will be displayed. If the time base is set for 1 μsec/cm, approximately two cycles will be displayed. If the time base is set at a frequency higher than the input frequency, only a part of the input signal will be displayed.

In the basic oscilloscope, the sweep generator runs continuously (is free-running), while in the more elaborate oscilloscopes it is normally turned off. In the type of oscilloscope under discussion (front panel shown in Fig. 7-13) the sweep generator requires a signal from some source to start or trigger it. This type of oscilloscope is called a *triggered oscilloscope*. The triggered oscilloscope permits more accurate time measurements to be made and gives a more stable presentation.

Front panel controls on this type of instrument are provided to permit the selection of the source and polarity of the trigger signal and the amplitude. In addition, provision is usually made to adjust the sweep generator to free-run. These controls are the TRIGGERING and LEVEL controls. The triggering control is a multiposition switch, and the level control is a potentiometer.

The triggering control chooses the source and polarity of the trigger signal. The source may be LINE, INTERNAL, or EXTERNAL, and the polarity selected may be negative ($-$) or positive ($+$). When LINE is selected, the power line frequency (e.g., 60 Hz) is the trigger frequency. When internal (INT) is selected, part of the input signal is tapped off from the vertical-deflection amplifiers and the input signal frequency is also the trigger frequency. In external (EXT), a front-panel binding post is connected through the switch to the sweep-generator circuit and a signal from any external source that is appropriate may be connected to the binding post.

The LEVEL control sets the amplitude level that the triggering signal must exceed to start the sweep generator. With the stability control misadjusted or with no signal available for triggering, the oscilloscope screen will be blank. When the level control is fully counterclockwise, it is in a position marked AUTO. In this position, the sweep generator will be free-running. As the level control is turned slowly clockwise, the trace will disappear from the cathode-ray tube. Continue turning the control clockwise until the trace reappears, and then stop turning. At this point, there will be a stable display on the screen.

The triggering and level controls are used to synchronize the sweep generator with the input signal. This gives a stationary waveform display. If they are unsynchronized, the pattern tends to jitter or move across the screen, making observation difficult.

Most oscilloscopes provide a test signal to a front-panel connector. This signal may be a few volts ac tapped from the power transformer, or it may be an accurately calibrated square wave. There may be just one panel connector with only one amplitude of voltage available, or there may be several connectors, each providing a different amplitude signal. Some models provide one connector with a switch to select any one of a wide range of amplitudes. The connector or connectors may be labeled TEST SIGNAL, LINE, or VOLTAGE CALIBRATOR. The oscilloscope under discussion here uses three jacks labeled VOLTAGE CALIB, with the output amplitude of a square wave marked over each jack. The voltages available are 5 volts, 0.5 volts, and 0.05 volts peak-to-peak. The voltage calibrator provides a known reference voltage for checking the calibration of the VOLTS/CM control. For example, suppose the VOLTS/CM control is set to the 1 VOLT/CM position and the VARIABLE control is in the calibrated position. With the five-volt calibrated waveform connected to the vertical input terminal, the presentation should be five centimeters in height, as shown in Fig. 7-22.

Most oscilloscopes provide a means of connecting an external signal to the horizontal-deflection amplifiers in place of the sawtooth from the sweep generator. In the oscilloscope shown in Fig. 7-13, when the TIME/CM control is rotated fully clockwise to the position marked EXT (external), the sweep generator will be disconnected from the horizontal-deflection amplifiers, and a front-panel binding post (HORIZ) will be connected to the input of the amplifiers. A signal may now be connected to this binding post. The VARIABLE TIME/CM control becomes an amplitude (gain) control to provide a means of controlling the width of the trace when a signal is applied. With no signal applied to the vertical or horizontal input connectors, a small, stationary dot will be present on the cathode-ray tube. However, this dot should not be allowed to remain on the CRT. Either a signal should be applied to one of the inputs, or the INTENSITY control should be turned counterclockwise until the dot disappears. Once a signal is applied, the intensity may be turned back up to view

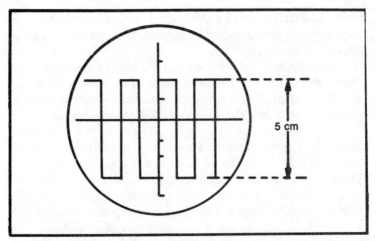

Fig. 7-22. The voltage calibrator provides a known reference voltage for checking alignment of the oscilloscope. Here, a five-volt input to the vertical terminal creates a waveform that is five centimeters in height.

the trace. If this dot is allowed to remain on the CRT, it will damage the chemical coating and necessitate the replacement of the CRT.

The horizontal input provides a means of connecting a second signal for the purpose of comparing phase or frequency differences with respect to the signal applied at the vertical input. The resultant display on the CRT is called a *Lissajous figure*.

Figure 7-23 shows the resultant pattern and how it was produced when two sine waves of the same frequency, but differing in phase by 90 degrees, are applied to the vertical and horizontal inputs. Notice

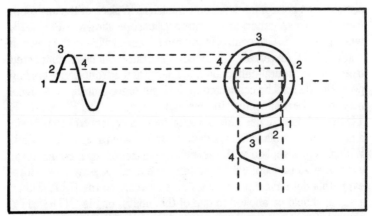

Fig. 7-23. Resulting pattern formed by two sine waves 90 degrees out of phase and of the same frequency.

194

that the pattern is a circle. Figure 7-24 shows the patterns that will result from various sine waves differing in phase. For accurate measurements, both signals should be the same amplitude on the screen. This is accomplished by applying only one signal at a time and adjusting the respective gain control.

Figure 7-25 shows the pattern that will result from applying a sine wave to the vertical input at twice the frequency of a sine wave applied to the horizontal input. To determine the actual frequency, one of the input frequencies must be known. Otherwise, the only information that will be obtained is the ratio of the two frequencies.

Using Lissajous figures to determine an unknown frequency is accomplished by first establishing a ratio. This is done by counting the number of times the pattern touches a horizontal line, A, in Fig. 7-25, and the number of times the pattern touches a vertical line, B, in the same figure. In this example, the ratio is 2:1 and represents the ratio of the vertical frequency to the horizontal frequency. If the horizontal frequency is known to be 1 kHz, then the vertical frequency is 2 kHz. The number of patterns that may be obtained is quite varied and ranges from the simple to the complex.

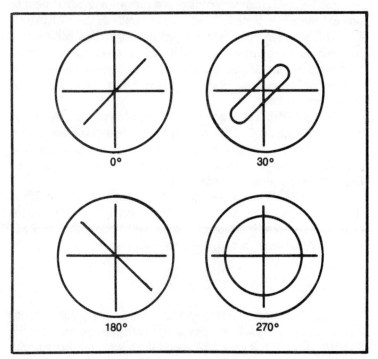

Fig. 7-24. Various patterns will result from different phase relationships.

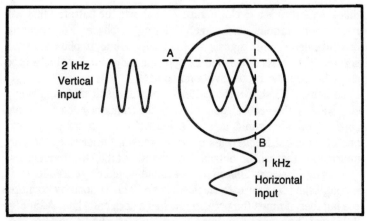

Fig. 7-25. Pattern of cathode-ray tube when the vertical input frequency is twice that of the horizontal input.

The ILLUM (illuminate) control turns on several small light bulbs mounted around the edge of the graticule, making the lines visible so that measurements of the displayed signal can be made easily.

The DC BAL (balance) control is a screwdriver adjustment to prevent the whole trace from shifting vertically as the VARIABLE VOLTS/CM control is turned through its range. This control may be mounted on the front panel or it may be a service adjustment located inside the oscilloscope. This control, once set, only requires adjustment at periodic intervals rather than every time the oscilloscope is used.

The STABILITY control is an adjustment that works in conjunction with the LEVEL control to provide a stable display. On the model under discussion, it is a screwdriver adjustment and normally requires no adjustment by the user. On more sophisticated oscilloscopes, it is a knob adjustment.

The 5× MAG (magnifier) essentially expands the sweep. When the magnifier is turned on, it increases the sweep speed by a factor of five. If the TIME/CM control were set at one millisecond and the 5× MAG were turned, the new sweep speed would be:

$$\frac{1 \text{ msec}}{5} = 0.2 \text{ msec.}$$

The magnifier permits you to examine more closely one cycle or a group of cycles rapidly without necessitating the resetting of the TIME/CM control. Figure 7-26 shows the visual presentation with the magnifier off and then turned on. On the oscilloscope being dis-

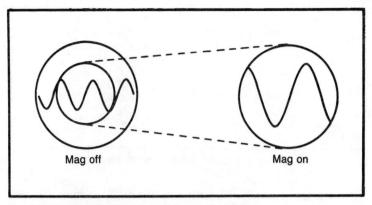

Fig. 7-26. Result of using the oscilloscope magnifier.

cussed, the 5× MAG is turned on by pulling the horizontal-positioning control straight out.

OPERATION OF THE OSCILLOSCOPE

The oscilloscope is one of many tools used for troubleshooting. Before using this instrument, a check should be made to verify that it is in good operating condition. If the oscilloscope is faulty, it will give false readings and you will find yourself wasting many hours trying to find a problem where none exists.

You should also be familiar with the operation and limitations of the oscilloscope being used. Misadjusting controls will cause false readings to be obtained. Attempting to take measurements in a circuit operating beyond the capabilities of the oscilloscope will also give false readings and may even result in the instrument being damaged. When using an oscilloscope for the first time, consult the manual for that particular instrument to determine the correct operating procedure and any limitations.

Since the oscilloscope provides a visual display which is affected by the manipulation of the controls and usually has its own built-in test signal source, its operating condition can be checked rapidly.

For the initial setup of the oscilloscope under discussion in this chapter (Fig. 7-13), the controls should be set as listed in Fig. 7-27. Allow approximately two minutes for the instrument to warm up. After the warm-up period, a trace should be visible. Turn the positioning controls one at a time to see if the trace can be moved to the edges of the CRT face, and then center the trace. Adjust the intensity control for comfortable viewing and the focus control for a thin, well-defined line.

Position, Focus, and	:	in the center
Intensity controls		of their range
		of rotation
TRIGGERING	:	INT +
LEVEL	:	AUTO
TIME/CM	:	1 msec
VARIABLE	:	CAL
POWER	:	ON

Fig. 7-27. Control position for initial setup of the oscilloscope.

Turn the VOLTS/CM control to the 0.02 position and the VARIABLE control to CAL. Connect the 0.05-volt output from the voltage calibrator to the vertical input. Turn the LEVEL control from the AUTO position. As this is done, the trace will disappear. Continue turning until the trace reappears. The display should be a stable display of several cycles of a square wave, 2.5 centimeters in height. Disconnect the lead from the voltage calibrator. The instrument is now ready for use.

Before connecting a signal to the input terminal, turn the input attenuator (VOLTS/CM) to the highest attenuation setting (10 volts/cm). Connect the input signal and rotate the attenuator control until the display occupies approximately two-thirds of the screen height. Adjust the time base and level controls to select the desired number of cycles of the input and stabilize the display.

OSCILLOSCOPE APPLICATIONS

An oscilloscope can be used in a myriad of applications and should be considered as almost essential in any electronics workshop. It may be used in servicing such devices as televisions, stereo equipment, radios, medical instruments, and automobiles and is widely used in industry for the testing of large equipment. Some of the basic ways in which an oscilloscope can be used will be covered to give the reader an idea of just what is possible with this multipurpose instrument.

Analysis of Waveforms

Sine waves can be used to form all complex waveforms. To see this, refer to Fig. 7-28. Here, a square wave has been built up from sine waves, each of which has an amplitude of a specific nature and a harmonic relation to the fundamental. Notice also that all the

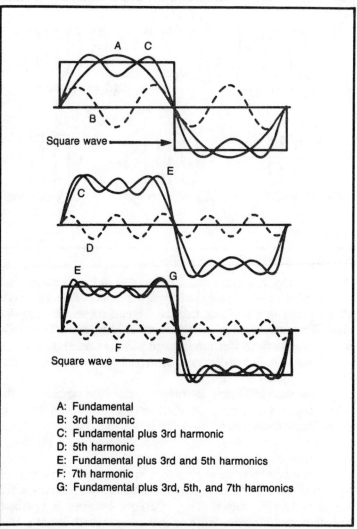

A: Fundamental
B: 3rd harmonic
C: Fundamental plus 3rd harmonic
D: 5th harmonic
E: Fundamental plus 3rd and 5th harmonics
F: 7th harmonic
G: Fundamental plus 3rd, 5th, and 7th harmonics

Fig. 7-28. Sine waves can be used to form all complex waves.

harmonics of a square wave are in phase. Also note that only odd harmonics make up the composition of a square wave. This is one of the characteristics of symmetrical waveforms. It may be helpful to analyze some examples of distorted square waveforms. To illustrate this, some of the higher harmonics of an ideal square wave would be removed if it were processed through a low-pass filter. Figure 7-29 shows some distorted waveforms. Note that the corners are rounded. This is due to the removal of the higher harmonics and

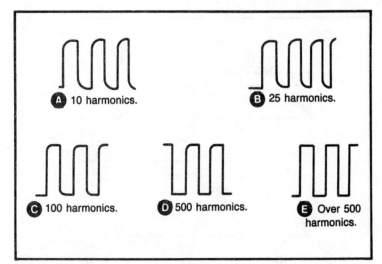

Fig. 7-29. Examples of distorted waveforms.

occurs because more harmonics are needed to form a square corner. Another phenomenon which occurs and can cause distortion is shown in this figure. Notice that the diagonal corners are rounded also. This is attributed to what is known as *phase shift*. To explain, if the low-pass filter which the square wave is passed through has zero phase shift, this will cause all corners of the distorted waveform to appear equally rounded.

It is also important to mention that the setting of the oscilloscope's gain control can affect the square wave to some degree. This is illustrated in Fig. 7-30. Here, both displays are of the same square-wave voltage, but the vertical gain control in the display on the right has been adjusted higher.

A sawtooth waveform is shown in Fig. 7-31. This type of waveform can be formed from both even and odd harmonics which are in phase, as shown. The difference between a sawtooth waveform and a square wave is that the sawtooth is not a true symmetrical waveform as is the square wave. Generally speaking, waveforms which are not symmetrical will contain both even and odd harmonics, although there are some which are formed from even harmonics only. One example of this is the full-wave rectifier shown in Fig. 7-32, which contains only even harmonics.

Another helpful fact to know is that all ac waveforms have both equal positive and negative areas above and below the zero-volt or beam resting level. Referring to the pulse waveform shown in Fig. 7-33, this waveform has this property. Conversely, any pulsating dc waveform or ac waveform which has a dc component will have unequal positive and negative areas above and below the zero-voltage level, as shown in Fig. 7-34. On an ac oscilloscope, all three waveforms will appear exactly the same due to the rejection of the dc component. These same three waveforms, however, will appear at varying vertical levels in the display. This makes it possible to measure the dc component and the peak-to-peak value of the ac component as well.

The exponential waveform commonly occurs during the charge and discharge of a capacitor in an RC circuit. This is shown in Fig. 7-35. Both the charge and discharge exponential are the same, except the discharge exponential is upside down. The time constant of an RC circuit is equal to the resistance in ohms times the capacitance in farads. The answer arrived at as a result of this equation will provide the user the amount of time (in seconds) which it takes for the voltage of the capacitor to rise to 63% of its final potential, or the time (in seconds) for the capacitor voltage to fall to 37% of its maximum potential. This is shown in Fig. 7-36.

A frequency-response curve is shown in Fig. 7-37. This is a very important type of waveform, particularly with regard to the repair and testing of radios and televisions. A frequency-response curve measures the signal output voltage over frequency bands. In order to produce a frequency-response curve, the user applies an FM signal to the tuned circuits being tested, as shown in Fig. 7-38. The pattern which appears on the display of the oscilloscope will indicate markers,

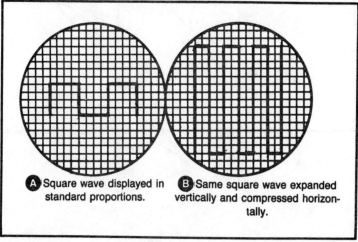

A Square wave displayed in standard proportions.

B Same square wave expanded vertically and compressed horizontally.

Fig. 7-30. A basic square wave can be greatly affected by the oscilloscope's gain controls.

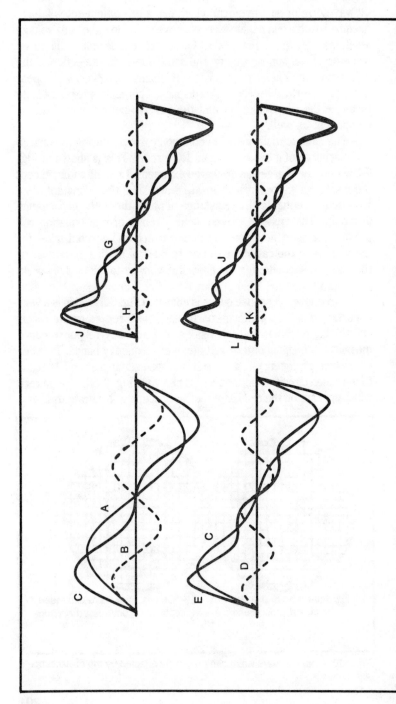

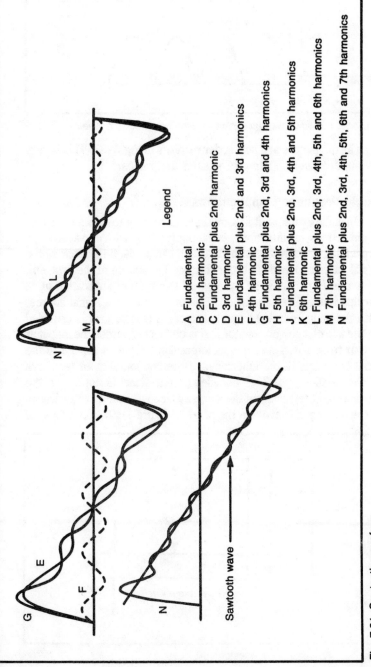

Legend

A Fundamental
B 2nd harmonic
C Fundamental plus 2nd harmonic
D 3rd harmonic
E Fundamental plus 2nd and 3rd harmonics
F 4th harmonic
G Fundamental plus 2nd, 3rd and 4th harmonics
H 5th harmonic
J Fundamental plus 2nd, 3rd, 4th and 5th harmonics
K 6th harmonic
L Fundamental plus 2nd, 3rd, 4th, 5th and 6th harmonics
M 7th harmonic
N Fundamental plus 2nd, 3rd, 4th, 5th, 6th and 7th harmonics

Sawtooth wave

Fig. 7-31. Sawtooth waveform.

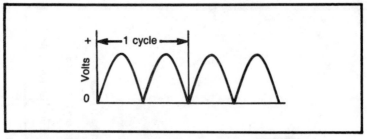

Fig. 7-32. Full-wave rectifier output as seen on an oscilloscope.

such as the picture and subcarrier markers in Fig. 7-37. This is very helpful in troubleshooting televisions and radios.

Audio-Frequency Applications

In audio systems, it is sometimes necessary to locate the cause of distortions, malfunctions, or no operation at all. An oscilloscope is useful in any of these situations and can be used to check all the signal paths and leads in order to trace the source of the problem. Shown in Fig. 7-39 is an example of this type of application of an oscilloscope. Here, an audio oscillator is used as a signal source, with the scope serving as an indicator. A normal load is provided by the use of a power resistor that is connected across the amplifier output terminals. Now, the audio-oscillator signal is applied to the input terminals of the amplifier and the horizontal input terminals of the oscilloscope. The amplifier output signal is applied to the vertical input terminals of the scope, which will result in a Lissajous pattern being displayed on the screen. Figure 7-40 shows some of

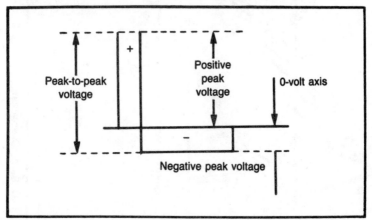

Fig. 7-33. A typical pulse waveform.

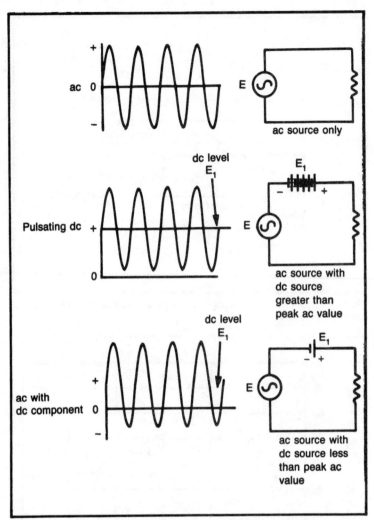

Fig. 7-34. Pulsating waveforms will have unequal positive and negative areas.

the patterns which may occur and evaluates them individually. If the pattern shown in A is displayed, this indicates that the amplifier is operating properly. The pattern in B means that there is an overload somewhere in the circuit. If the overload is extreme, the pattern will appear as that shown in C. The display in D is an indication that phase shift is occurring. Although phase shift in itself is not a problem, particularly in amplifiers, if it occurs in conjunction with an overload, as shown in E and F, it is an indication of other types of problems that must be corrected. The pattern shown in G is most

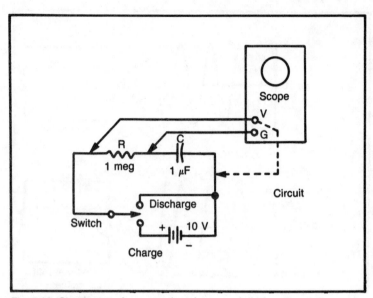

Fig. 7-35. Circuit setup for measuring charge and discharge waveforms of a capacitor in an RC circuit.

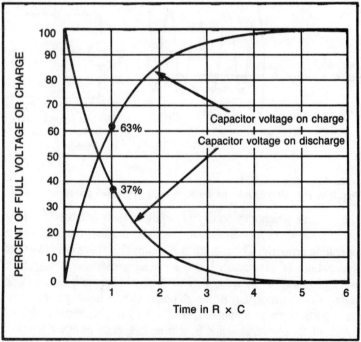

Fig. 7-36. A typical readout of charge and discharge time in an RC circuit.

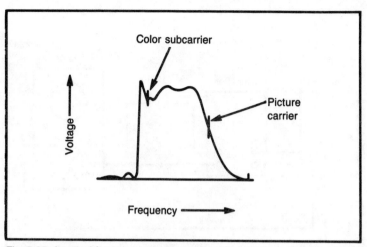

Fig. 7-37. A typical frequency-response curve of a television signal as shown on an oscilloscope.

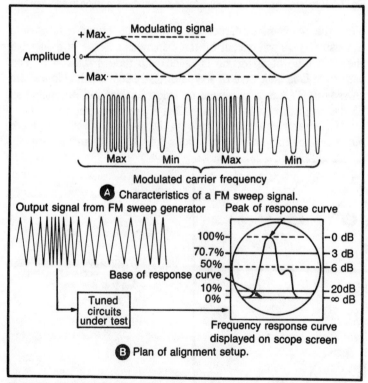

Fig. 7-38. When testing for frequency response, an FM signal is applied to the tuned circuits.

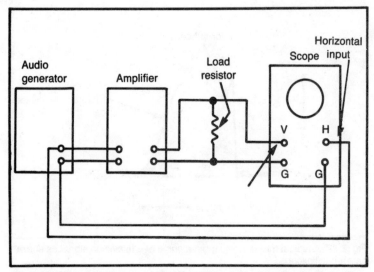

Fig. 7-39. Typical oscilloscope setup for testing audio distortion in an amplifier.

often the result of incorrect bias voltage on either a tube or a transistor. This will appear on the display as a curve of a diagonal line; this type of distortion is called *amplitude nonlinearity*. The pattern in G is indicative of a type of distortion which is known as *crossover*. This is also caused by incorrect bias but normally can be traced to the push-pull stage of the amplifier circuit.

The examples provided here on the use of an oscilloscope are but a few of the many applications for this instrument. Anyone

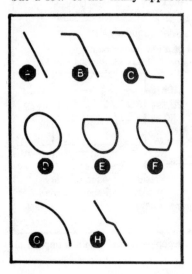

Fig. 7-40. Some of the resulting patterns that may occur when using the previous test system.

purchasing an oscilloscope will receive a complete instruction manual which discusses the capabilities of the instrument in much more detail. Diagrams and pictures are also often provided in order for the user to see exactly where in a circuit to connect the scope to test for different values.

THE DUAL-TRACE OSCILLOSCOPE

The information contained thus far in this chapter has dealt specifically with the basic oscilloscope and its operation. A variation on the basic instrument is the *dual-trace oscilloscope,* which permits the simultaneous viewing of two independent signal sources on a single cathode-ray tube screen. This type of operation affords an accurate method of making amplitude, phase, or time-displacement comparisons and measurements between two signals at the same time. A dual-trace oscilloscope should not be confused with a *dual-beam oscilloscope.* Dual-beam oscilloscopes are those which produce two separate electron beams on a single scope and which can be individually or jointly controlled. Dual trace refers to a single beam in a CRT that is shared by two channels.

There are two methods by which the single beam can be shared. The first method of obtaining a dual trace is called the *chop mode.* Shown in Fig. 7-41 is a simplified block diagram of the dual-trace section utilizing the chop mode. The dc output reference voltage on each of the amplifiers is adjustable. Thus, the beam will be deflected by different amounts on each channel if the voltage reference is different at each amplifier output. The output voltage

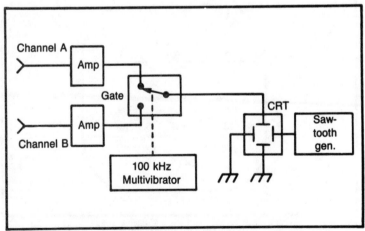

Fig. 7-41. Block diagram of a portion of the dual-trace oscilloscope.

from each amplifier is applied to the deflection plates through the gate. The gate is actually an electronic switch. In this application, it is commonly referred to as a beam switch. In the chop mode the switch is controlled by a high-frequency multivibrator. That is, the gate selects one channel's output and then the other at a high frequency rate, which is 100 kHz in most oscilloscopes. Because the switching time is very short in a quality scope, the resulting display is two horizontal dashed lines. This is shown in Fig. 7-42. Line A is the output of one channel, while line B is the output of the other. The trace goes from left to right due to a sawtooth waveform applied to the horizontal plates. A more detailed analysis shows that the beam moves from 1 to 2 while the gate is connected to the output from one channel. Then, when the gate samples the output of the second channel during time 3-4 (assuming channel 2 is at a different voltage reference), the beam is at a different vertical location. The beam continues in the sequence 5-6, 7-8, 9-10, and 11-12 through the rest of one horizontal sweep. When the chopping frequency is much higher than the horizontal sweep frequency, the number of dashes will be very large. For example, if the chopping occurs at 100 kHz and the sweep frequency is 1 kHz, then each horizontal line would be comprised of 100 dashes. This display would then look like a series of closely spaced dots, as shown in Fig. 7-43. As the sweep frequency becomes lower relative to the chopping frequency, the display will show two apparently continuous traces. Therefore, the chop mode is used at low sweep rates (low time per

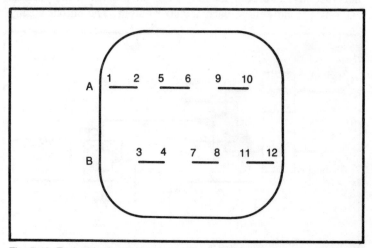

Fig. 7-42. Typical horizontal traces on the A and B channels of a dual-trace oscilloscope.

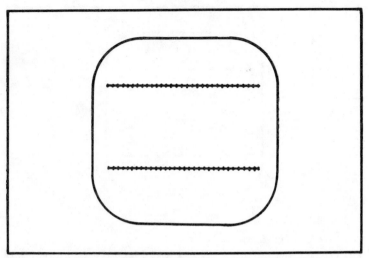

Fig. 7-43. Dual-trace display of the horizontal lines using a chopping frequency of 100 kHz and a sweep of 1 kHz.

division settings). When signals are applied to the channel amplifiers, the outputs are changed in accordance with the input signal. The resultant pattern on the screen gives a time-base presentation of the signal of each channel as shown in Fig. 7-44. Part A shows the chopped traces without signals; B depicted the signals into the two channels; and C is the resulting display.

The second method (alternate mode) of obtaining a dual-trace function uses the technique of gating between sweeps. This is shown in Fig. 7-45. The gate samples one channel for one complete sweep and the other channel for the next complete sweep. The gate selection is controlled by the sweep circuitry. At low sweep speeds, one trace begins to fade while the other channel is being gated. Consequently, this mode is not used for slow sweep speeds. Because the chop mode will not operate satisfactorily at high speeds and the alternate mode is deficient at low speeds, both are used on dual-trace oscilloscopes to complement each other.

DUAL-TRACE OSCILLOSCOPE CONTROLS

Because the settings of the controls on the front panel change the operation of an oscilloscope, it is necessary for you to have an understanding of the effects of these controls. The front panel in Fig. 7-46 will be referred to throughout this discussion.

The ON-OFF switch controls the application of primary power to the oscilloscope. On many oscilloscopes, this switch is part of

211

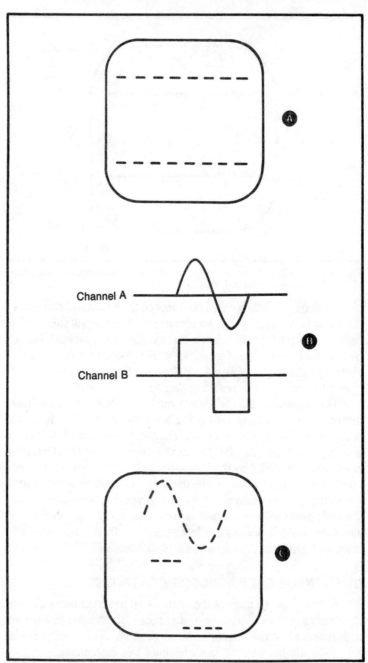

Fig. 7-44. Time-base presentation of the signal of each channel on a dual-trace oscilloscope.

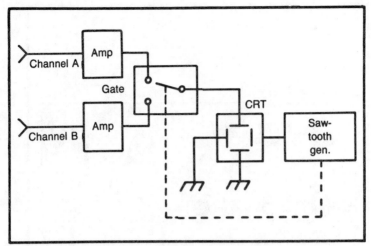

Fig. 7-45. The alternate mode of obtaining a dual trace.

another control, such as the INTENSITY control or GRATICULE control. (The *graticule* is the overlying grid on the face of the cathode-ray tube.) On this scope the ON-OFF switch is a part of the GRATICULE control. Some scopes have a lamp near the ON-OFF switch that illuminates to indicate the power-on condition.

The GRATICULE control varies the graticule illumination intensity. The best setting will vary, depending upon ambient light conditions and the specific use of the oscilloscope.

Adjustment of the INTENSITY or BRILLIANCE control varies the brightness of the spot or trace. Internally, this control varies the bias of the cathode-ray tube.

A small clear spot or trace is obtained by setting the FOCUS control, while a trace of uniform width across the entire screen is dependent upon the setting of the ASTIGMATISM (ASTIG) control. Because the INTENSITY, FOCUS, and ASTIGMATISM controls interact, they must be adjusted in relation to one another to obtain a uniform trace with maximum resolution at the desired intensity.

The input coupling switch determines the type of voltage that will be applied to the vertical channel. Figure 7-47 shows the basic coupling networks in AC, DC, and GROUND positions.

The capacitor in the AC position blocks the dc component of the input, so only a change in voltage is coupled to the vertical deflection circuits. The input to the vertical channel is directly coupled in the DC position. A third position (GND) is used to ground the vertical amplifier input and open the input signal path between the input jack and the amplifiers.

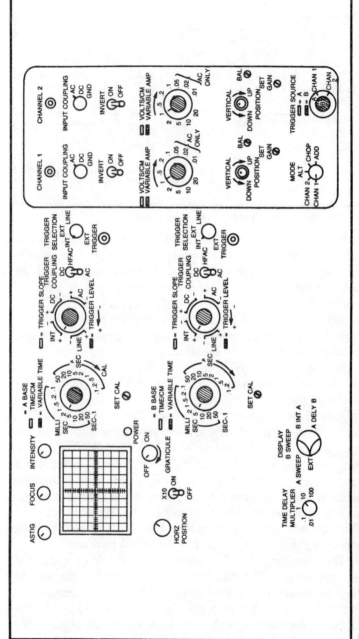

Fig. 7-46. Controls on a typical dual-trace oscilloscope.

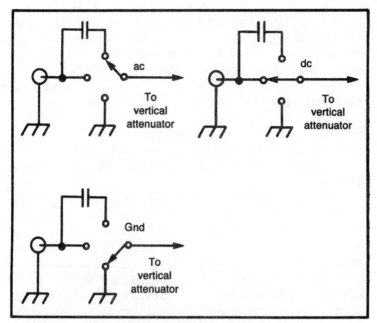

Fig. 7-47. Some vertical channel coupling circuits.

The VOLTS/CM switch selects different settings on an input attenuator network. When the setting of the VOLTS/CM switch is changed, the amount of attenuation of the input signal is varied, and the amplitude of the display is changed. The gain of the vertical amplifiers is adjusted during calibration so that each position of the VOLTS/CM switch will provide a vertical deflection of a specified amount for a given signal strength. For example, in the one-volt-per-centimeter position, each centimeter that the display covers vertically represents one volt (the graticule is usually divided into divisions of one centimeter).

The VARIABLE AMPLITUDE control (generally a part of the VOLTS/CM switch) provides another means of varying the amplitude of the display. This control varies the gain of the vertical amplifiers. If the VARIABLE control is not fully clockwise, the vertical deflection will be less than what is specified by the VOLTS/CM position.

The VERTICAL POSITION (SHIFT) control affects the vertical location of the display. This is accomplished by varying the dc potential between the vertical deflection plates.

The BALANCE (BAL) control is normally a screwdriver adjustment, which, when properly adjusted, prevents shifting of the trace as the VARIABLE AMPLITUDE control is adjusted.

The SET GAIN control is a screwdriver adjustment that varies the gain of the vertical amplifiers. When it is properly adjusted with the VOLTS/CM switch in the CAL position, the amount of vertical deflection will be correct for the other VOLTS/CM settings.

If a ×10 switch is incorporated, it switches in an additional amplifier stage that has a gain of ten. This increases the amplitude of the signal applied to the vertical deflection plates by ten. This frequency response will usually be much less in this position and should not be used unless absolutely necessary.

An INVERT switch reverses the phase of the voltage applied between the vertical deflection plates. This can be used to superimpose two signals that are actually 180 degrees out of phase.

The MODE switch determines the channel that is being gated to the vertical deflection plates. Usually, the following modes are available: channel 1 only (CHAN 1); channel 2 only (CHAN 2); alternate (ALT); CHOP; or ADD. In the channel 1 only mode, the beam switch remains connected to channel 1 and the oscilloscope behaves essentially as a single-trace scope. The channel 2 only mode applies the voltage on channel 2 only to the vertical-deflection system. When switched to the ALTERNATE mode, the oscilloscope requires high horizontal sweep speeds for signal comparison. One channel will appear on the screen during one sweep and the other channel output will appear during the other sweep. For low sweep speeds, the CHOP mode, which presents both channels during the same sweep, is used. In the ADD mode, the signal voltage on channel 1 is algebraically added to the signal voltage on channel 2. Consequently, a single trace which represents the sum of the two channels is displayed.

The TRIGGER SOURCE selector control allows either channel input to trigger either sweep trace. A common example of the use of this control is the selection of channel 1 or channel 2 as the trigger or synchronizing source for the horizontal sweep. However, some oscilloscopes have two independent horizontal sweeps which can be triggered in any combination from channel 1 and channel 2; i.e., sweep A, as shown in Fig. 7-46, can be triggered by either channel and sweep B can be triggered by either channel. In other oscilloscopes, a composite signal of A and B can be selected to trigger the sweep circuits.

The setting of the TRIGGER SELECTION switch to the INTERNAL (INT) position makes it possible to obtain the trigger pulse internally from the vertical amplifiers. The channel that does the triggering is selected by the setting of the TRIGGER SOURCE selector just described. By switching the trigger selection switch to EXTERNAL (EXT), the trigger can be obtained externally from whatever source is being applied through the external trigger jack. Some units, like this one, also have a LINE position which outputs a low-amplitude 60-Hz line voltage. This can also be used for triggering. These positions are sometimes called *trigger modes*.

The TRIGGER COUPLING switch usually has three positions, DC, AC, and HIGH FREQUENCY AC (HFAC). The dc position provides direct coupling to the trigger circuits, whereas the AC position incorporates a coupling capacitor to block any dc component. The high-frequency ac (HFAC) position has a high-pass filter which passes only those trigger signals that are above a certain frequency.

The position of the TRIGGER SLOPE switch determines whether the sawtooth sweep will be initiated on the positive-going portion or on the negative-going portion of the signal to be displayed. In Fig. 7-48, the positive-going portion occurs from time A to B, and the negative portion occurs from time B to C.

The TRIGGER LEVEL determines the voltage level required to trigger the sweep. For example, in the internal trigger mode, the trigger is obtained from the signal to be displayed. Therefore, the setting of the trigger level will determine the point of the input waveform that will be displayed at the start of the sweep.

Figure 7-49 shows some of the displays for one channel that will be obtained for different trigger levels and trigger-slope settings. The level is zero and the slope is positive in A, while B also shows a zero level but a negative slope selection. Diagram C shows the

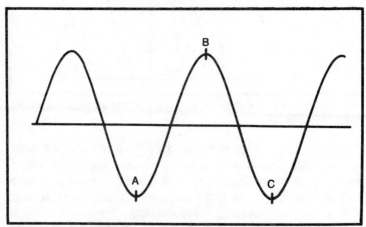

Fig. 7-48. The trigger slope switch will determine whether the sawtooth sweep will occur on the positive or negative-going portion of the display signal.

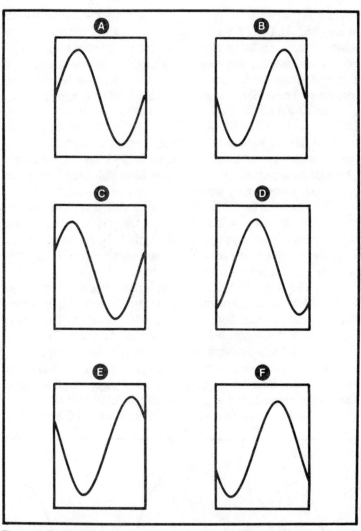

Fig. 7-49. Displays for one channel that result from different trigger levels and trigger-slope settings.

effects of a positive trigger-level setting and a positive trigger-slope setting, while D displays a negative trigger-level setting with a positive trigger-slope setting. Diagrams E and F have negative slope settings. The difference is that E has a positive trigger-level setting, while F has a negative trigger-level setting.

The automatic function of the trigger circuitry allows a free-running trace without a trigger signal. But when a trigger signal is applied, the circuit reverts to the triggered mode of operation and the sweep no longer free-runs. This provides a trace when no signal is applied.

Synchronization is also used to cause a free-running condition without a trigger signal. Synchronization is not synonymous with triggering. Triggering refers to a certain condition which initiates an operation. Without this condition, the operation would not occur. In the case of the triggered sweep that was just presented, the sweep will not start until a trigger is applied, and each succeeding sweep must have a trigger before a sweep commences. *Synchronization* means that some operation or event is brought in step with a second operation. A sweep circuit which uses synchronization instead of triggering will cause a previously free-running sweep to be locked in step with the synchronizing signal. The trigger-level control setting must be increased until synchronization occurs. Until this time, an unstable pattern appears on the cathode-ray tube face.

The A TIME BASE circuitry produces a sawtooth waveform that is applied to the horizontal deflection plates on the cathode-ray tube in order to obtain a sweep. The rate of the sweep is changed by the TIME/CM switch. For a given setting of this switch and with the variable time control set to the CAL position, the movement of the beam from left to right across one centimeter (into which the graticule is divided) occurs in the time specified by the TIME/CM switch setting. The variable time control varies the rate of the sweep so that any sweep speed in between those set by the TIME/CM switch can be obtained. Most oscilloscopes have a B time-base circuitry which has a TIME/CM and variable time control which is similar to the A time-base circuitry. The horizontal position or X shift control adjusts the horizontal position of the trace for both the A and B time bases.

There is normally a screwdriver adjustment called SET CAL which varies the sweep rate. This is adjusted to obtain the exact time-per-centimeter sweep rate that is indicated by the TIME/CM setting.

Many oscilloscopes have a ×10 switch in the horizontal-amplifier section. Just as in the vertical section, this switch adds another amplifier. In the sweep mode, the sawtooth voltage change will be ten times greater than that required for a complete left to right deflection. As a result, only one-tenth of the entire sweep will be visible on the screen.

The heart of sweep circuit control is the DISPLAY switch. In the external (EXT) position of the switch, an external signal must be applied to the external horizontal input jack in order to obtain horizontal deflection. The relationship between the type of waveform applied to the external jack and the y input jack will determine the resultant display. Two examples are time-base presentation, like those obtained when using internal sweeping, and Lissajous patterns.

In the A sweep position of the display switch, internal sweeping occurs as determined by the A TIME/CM controls. Some oscilloscopes have a B SWEEP position on the display switch so that either the A sweep or the B sweep can be used. In many oscilloscopes, the B sweep is used only for the purpose of a B intensified by A and A delayed by B.

In the B intensified by A (B INT A) position of the display switch, the B sweep is applied to the horizontal-deflection plates and a rectangular pulse equal to the rise time of the A sweep is added to the unblanking pulse from the B sweep. This will cause the beam to be deflected at the B sweep rate and also intensified a greater amount during a time equal to the A sweep. The A sweep time should always be shorter than the B sweep time. The segment of the B sweep which will be intensified will be determined by the setting on the time-delay multiplier. The settings on the time-delay multiplier multiplied by the B sweep TIME/CM setting determines the amount of time that will elapse between the initiation of the B sweep and the initiation of the A sweep. An example is shown in Fig. 7-50.

Waveform A represents the A sweep, while waveform B represents the B sweep. The input signal to one of the vertical channels is represented by waveform C. Waveform D is the composite unblanking pulse, which is the addition of a pulse of a time duration equal to the rise time of waveform B and a pulse of a time duration equal to the rise time of waveform A. As mentioned, the start of waveform A is determined by the setting of the time-delay multiplier. Varying the setting of this control will change the portion of the input waveform that is intensified. Figure 7-49E shows the resultant display.

The A delayed by B (A DELAY B) mode display occurs only during the duration of the A sweep. Again, the B sweep must be longer than the A sweep, and the start of the A sweep in relation to the B sweep is determined by the time-delay multiplier. In this case, the beam is swept by the A sweep and only that segment of the input signal which occurs during the time of the A sweep will be displayed on the cathode-ray tube, because the CRT will be unblanked once during the A sweep. Figure 7-51 shows the typical waveforms when operating in the A delayed by B mode.

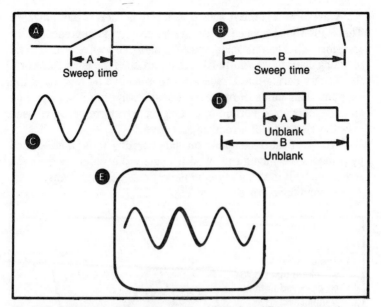

Fig. 7-50. Differences in the time duration of the A and B sweeps are the result of the settings of the time-delay multiplier multiplied by the B sweep TIME/CM settings.

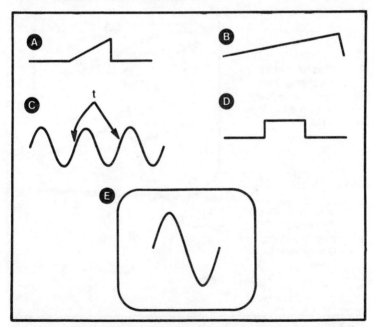

Fig. 7-51. Typical waveform when operating in the A delayed by B mode.

Waveform A is the A sweep, and waveform B is the B sweep. The input signal applied to a vertical channel is represented by waveform C. The unblanking pulse is shown by waveform D. The resulting display (waveform E) is that portion of the input waveform that occurs during the A sweep, as represented by time (t) in the diagram. This mode of operation is especially useful in displaying an extremely small portion of a complex waveform or for viewing only the rise time of a rectangular wave.

When in dual-trace operation and operating in the B intensified by A mode, the waveforms of both inputs will be intensified. In the A delayed by B mode, a segment of both channels will be displayed. These conditions are shown in Fig. 7-52.

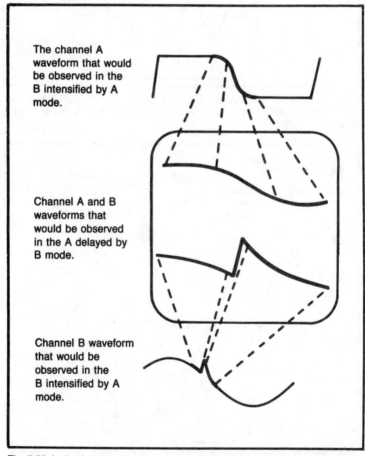

The channel A waveform that would be observed in the B intensified by A mode.

Channel A and B waveforms that would be observed in the A delayed by B mode.

Channel B waveform that would be observed in the B intensified by A mode.

Fig. 7-52. In the A delayed by B mode, a segment of both channels is displayed.

OPERATION

Applications of triggered oscilloscopes will be discussed here as well as the basic characteristics and operation of dual-trace oscilloscopes. Figure 7-53 shows a block diagram of a typical dual-trace oscilloscope without the power supplies. The circled letters in the block refer to the waveform designations to the right of the diagram. The waveform to be observed (waveform A) is fed into the channel A vertical-amplifier section. The VOLTS/DIV control sets the gain of the amplifier. In a similar manner, waveform B is applied to the channel B vertical-amplifier section. The beam switch determines the channel, as described earlier. Assuming that the chop mode is selected, the output of the beam switch is shown by waveform C. The waveform is fed through the delay line and vertical output amplifier to the vertical deflection plates of the cathode-ray

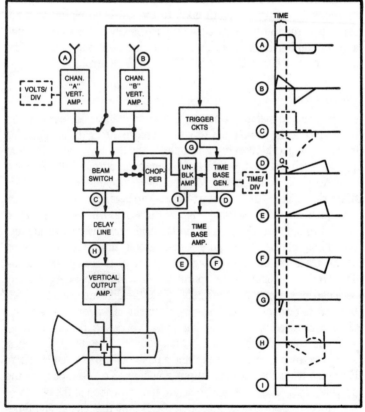

Fig. 7-53. Block diagram of a typical dual-trace oscilloscope.

tube. The purpose of the delay line will be discussed as well as the time-base generator.

Time-Base Generator

Just as in the single-trace oscilloscope, the time-base generator or sweep generator develops a sawtooth wave (waveform D) that provides the horizontal-deflection voltage. The rising or positive-going part of this sawtooth is linear. That is, the waveform rises through a given number of volts during each unit of time. This rate of rise is set by the TIME/DIV control. The sawtooth voltage is fed to the time-base amplifier. This amplifier includes a phase inverter so that the amplifier supplies two output sawtooth waveforms simultaneously. One of them is positive going (waveform E), while the other is negative going (waveform F). The positive-going sawtooth is applied to the right-hand horizontal-deflection plate and the negative-going sawtooth goes to the left-hand plate. As a result, the cathode-ray beam is swept horizontally to the right through a given number of graticule divisions during each unit of time. The sweep rate is controlled by the TIME/DIV control.

Delay Line

In order to maintain a stable display on the screen, each sweep must start at the same point on the waveform being displayed. This is accomplished by feeding a sample of the displayed waveform to a trigger circuit that produces a negative, output-voltage spike (waveform G) at a selected point on the displayed waveform. This triggering spike is used to start the run-up portion of the horizontal sweep. Because the leading edge of the waveform to be displayed is used to activate the trigger circuit, and since the triggering and unblanking operations require a measurable time, the actual start of the trace on the screen lags the start of the waveform to be displayed. This difference is approximately .14 microsecond in many oscilloscopes. Time interval t represents the difference in Fig. 7-53. In order to display the leading edge of the input waveforms, a delay, Q, is introduced by the delay line in the vertical-deflection channel after the point where the trigger is obtained. The delayed vertical signal is shown by waveform H, and a push-pull version of waveform H comes from the vertical-output amplifier. To re-emphasize the purpose of the delay line, it is to retard the application of the observed waveform to the vertical-deflection plates until the trigger and time-base circuits have had an opportunity to initiate the unblanking and horizontal-sweep operations. In this way, the entire waveform can be observed even though the leading edge of the waveform was used to trigger the horizontal sweep.

Unblanking

The unblanking operation is the application of a rectangular unblanking wave (waveform I in Fig. 7-53) to the grid of the cathode-ray tube. The duration of the positive part of this rectangular wave corresponds to the duration of the positive-going part of the time-base output (waveform D), so that the beam is switched on during its left-to-right excursion and is switched off during its right-to-left retrace.

For the input signals shown in Fig. 7-53, the waveform at many points will be different when operating in the alternate mode. Figure 7-54 depicts the waveforms in the alternate mode. Two sweeps are shown because it requires two sweeps to display the information

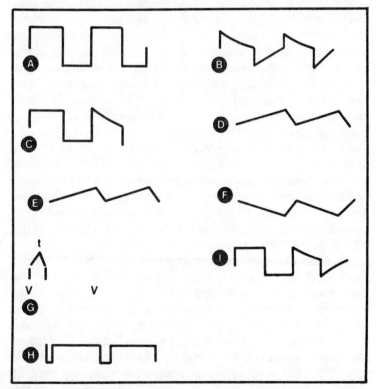

Fig. 7-54. Alternate mode waveforms.

of both channels in this mode. Notice that the unblanking voltage is removed in waveform I during the retrace. The trigger pulse (waveform G) is obtained from the same channel regardless of which input is being sampled. The input channel that triggers the sweep depends on the front-panel setting of the trigger selector. It can be seen from waveform C that during the first entire sweep, one channel is sampled and during the next sweep, the other channel is sampled.

This discussion on the basic dual-trace oscilloscope has covered the simplified operation of this instrument. Some of the operations have been simplified purposely due to the complex circuitry involved. As a result, there may be some slight deviations compared to other dual-trace oscilloscopes.

DUAL-TRACE OSCILLOSCOPE ACCESSORIES

The basic dual-trace oscilloscope has one gun assembly and two vertical channels. However, there may be many variations on this. The horizontal channels may vary somewhat from equipment to equipment. Some have one time-base circuit, while others have two; and these two are interdependent in some oscilloscopes, while others are independently controlled. Also, most modern general-purpose oscilloscopes have a modular construction. That is, most of the vertical circuitry is contained in a removable plug-in unit and most of the horizontal circuitry is contained in another plug-in unit. The main frame of the oscilloscope can then be adapted for many applications by designing a variety of plug-in assemblies. This modular feature provides much greater versatility in a single oscilloscope. For instance, the dual-trace plug-in models can be replaced with a semiconductor curve-tracer plug-in module if you want to analyze transistor characteristics. Other plug-in modules available with some oscilloscopes are high-gain, wide-bandwidth amplifiers, differential amplifiers, spectrum analyzers, physiological monitors, and other specialized units. Thus, the dual-trace capability is a function of the type of plug-in unit that is used.

In order to derive maximum usefulness from an oscilloscope, there must be a means of connecting the desired signal to the oscilloscope input. Aside from cable connections between an equipment output and the oscilloscope input, there are a variety of probes available which facilitate monitoring of signals at any point desired in a circuit. The more common types include: 1:1 probes, attenuation probes, and current probes. Each of these probes may be supplied with several different tips to allow for measurement of signals on any type of test point. Figure 7-55 shows some of the

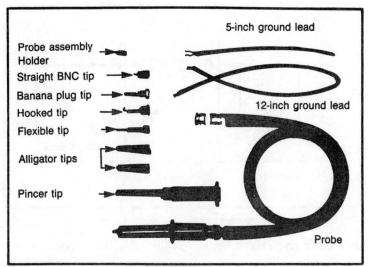

Fig. 7-55. Common probe tips used with dual-trace oscilloscopes.

more common probe tips. In choosing the probe to use for a particular measurement, you must consider such factors as circuit loading, signal amplitude, and scope sensitivity.

The 1:1 probe offers little or no attenuation of the signal under test and therefore is useful for the measurement of low-level signals. However, circuit loading with a 1:1 probe may be detrimental. The impedance at the probe tip is the same as the input impedance of the oscilloscope.

The attenuator probe has an internal high-value resistor in series with the probe tip. This gives the probe a higher input impedance than that of the oscilloscope, providing the capability to measure high-amplitude signals that would overdrive the vertical amplifier if connected directly to the oscilloscope. Figure 7-56 shows a schematic representation of a basic attenuation probe. The nine-megohm resistor in the probe and the one-megohm input resistor of the oscilloscope form a 10:1 voltage divider.

Because the probe resistor is in series, the oscilloscope input resistance when using the probe is ten megohms. Thus, using the attenuator probe with the oscilloscope will cause less circuit loading than with the 1:1 probe.

Before using an attenuator probe for measurement of high-frequency signals or for fast-rising waveforms, the probe compensating capacitor (C_1 in Fig. 7-56) must be adjusted according to instructions in the appropriate manual. Some probes will have an

227

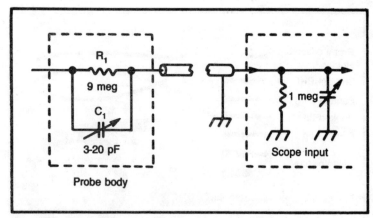

Fig. 7-56. Schematic diagram of basic attenuation probe.

impedance equalizer in the end of the cable that attaches to the oscilloscope. The impedance equalizer, when adjusted according to the manufacturer's instructions, assures proper impedance matching between probe and oscilloscope. An improperly adjusted impedance equalizer will result in erroneous measurements, especially when measuring high frequencies or fast-rising signals.

Current probes utilize the electromagnetic fields produced by a current. The probe is designed to be clamped around a conductor without having to disconnect the conductor. The current probe is electrically insulated from the conductor, but the magnetic fields about the conductor induce a potential in the current probe proportional to the current through the conductor. Thus, the vertical deflection of the oscilloscope display will be directly proportional to the current through the conductor.

SUMMARY

The oscilloscope is probably the most versatile instrument available today to the service technician. It is also one of the most misunderstood. Due to the multitude of controls, many people shy away from this instrument, feeling that it's just too complicated to operate. This is not the case. While many highly complex measurements may be easily made with the oscilloscope, simplistic checks may also be effected and with the same ease with which resistance measurements are made when using the ohmmeter.

The oscilloscope is included as a major part of nearly every test bench, but even the experienced technician may not utilize it to its best advantage. The oscilloscope is really many instruments in one, but so is the multimeter. The oscilloscope displays its readings on a cathode-ray tube, while the multimeter uses an analog or digital meter face. By understanding the oscilloscope as well as the multimeter, many test procedures can be conducted by even beginning service technicians more efficiently, more accurately, and in a shorter period of time.

Chapter 8

Building Your Own Test Instruments

Many electronic enthusiasts take great delight in building their own circuits rather than resorting to buying manufactured versions from the many companies that offer them. Home building is to be encouraged. This process leads to a better understanding of exactly how many different pieces of electronic equipment function and also makes home repairs that much simpler.

Electronic test instruments may also be built in the home shop, and a great deal of savings can be had by going this route. There are some test instruments which do not readily lend themselves to this type of construction due to the complexities involved, but many can be built in a few hours in the average home shop.

One compromise between building from scratch and purchasing commercially manufactured equipment is to build commercial kits. Heathkit is the best known manufacturer of electronic kits in the world today and they offer an especially attractive line of electronic test instruments. Some of these are highly specialized in nature, while others are designed for the person who is starting his first electronics test bench. I have assembled many Heathkit products which range from the simplest to the most complex. In every case, the assembly instructions have been very clear and concise. This is of the utmost importance when assembling any kit. The Heath Company has been at this pursuit for many decades and can certainly be classified as the leader in the electronics kit field.

HEATHKIT 5280 SERIES

The Heath Company offers a kit series of test instruments which is an excellent way to assemble a versatile test bench without spending a lot of money. The 5280 series of test instruments offers a choice of five different devices which include an audio generator, rf generator, high-performance multimeter, RCL bridge, and signal tracer. A special power supply is also available which can power all five instruments simultaneously.

Each instrument in the 5280 series is currently priced at $44.95 and includes all the components to assemble the complete device. This series is designed for beginning hobbyists, students, or service technicians. When all five instruments are purchased and assembled, you will be equipped to work on many types of electronic and electrical equipment. This series is designed with low cost in mind, but quality has not been ignored. They are excellent instruments and should serve the average bench technician for many years to come.

MULTIMETER

The first instrument in the 5280 series is the Heathkit IM-5284 High-Performance Solid-State Multimeter. This instrument gives you four different functions and a large, easy-to-read panel meter. It is capable of taking ac and dc voltage measurements up to one kilovolt and dc current measurements up to one ampere full scale. The ohmmeter function is divided into four ranges: ×1 ohm, ×100 ohms, ×10 kilohms, and ×1 megohms. The kit is complete and includes a high-impact molded case. The instrument is powered by two 9V dc batteries and one 1.5-volt C cell. The optional power supply will eliminate the need for these batteries. Figure 8-1 provides a table of specifications for the IM-5284 Multimeter.

RF GENERATOR

The second instrument in this series is the rf generator (called the Rf Oscillator by Heathkit). This unit includes probes and is designed for use in aligning the tuned stages of AM, FM, and television receivers. Labeled the IG-5280, this rf generator produces an output which is divided into five different bands of from 310 kHz to 110 MHz. It also features an extra 100 to 200 MHz band of calibrated harmonic frequencies. This generator is provided with its own internal, 1000-Hz audio output. The signal is available by means of a front panel jack and is ideal for tracing and isolating certain defects

IM-5284 SPECIFICATIONS: Dc Voltmeter: Ranges: 0-1, 0-10, 0-100, 0-1000 volts, full scale. **Accuracy:** ±3% of full scale. **Ac Voltmeter: Ranges:** 0-1, 0-10, 0-100, 0-1000 volts, full scale. **Accuracy:** ±5% of full scale. **Dc Milliammeter: Ranges:** 0-1, 0-10, 0-100, 0-1000 mA, full scale. **Accuracy:** ±4% of full scale. **Dc Input Resistance:** 10 megohms. **Ac Input Resistance:** 1 megohm. **Dimensions:** 5.75″ H × 11″ W × 7.75″ D (14.61 × 27.94 × 19.69 cm).

Fig. 8-1. Table of specifications for the Heathkit IM-5284 Multimeter.

in receiver audio stages. It also serves as a source of internal AM modulation. The IG-5280 kit is complete in every way. All the builder need do is assemble it according to Heathkit instructions and provide two 9 Vdc batteries for power. Figure 8-2 shows a table of specifications for this test instrument.

AUDIO GENERATOR

The IG-5282 Sine/Square Wave Audio Oscillator is the third instrument in the 5280 series and is designed to be used in many different audio test applications. Like the rf generator, the output from the audio oscillator is in several different ranges. Frequency coverage is from 10 Hz to 100 kHz in either sine or square-wave mode, selectable in four ranges. This audio generator may be used as a signal source during signal tracing in electronic circuits. With the appropriate associated equipment, the sine-wave output may be used for audio stage gain and distortion analysis. The use of the audio generator in distortion measuring is discussed elsewhere in this book. If you need to determine frequency response, the square-wave

IG-5280 SPECIFICATIONS: Rf Output: Frequency Range: 310 kHz to 110 MHz, in five bands; 100 MHz to 220 MHz on harmonics. **Output Voltage:** Approximately 100 mV. **Internal Modulation:** 1 kHz. **Af Output: Frequency:** 1 kHz. **Output Voltage:** 2.0 Volts rms (open circuit). **Dimensions:** 5.75″ H × 11″ W × 7.75″ D (14.61 × 27.94 × 19.69 cm).

Fig. 8-2. Table of specifications for the Heathkit IB-5280 Rf Generator.

IG-5282 SPECIFICATIONS: Frequency Output: 10 Hz to 100 kHz, in four ranges. Sine Wave Output Voltage: 0-3 Volts rms. Square Wave Output Voltage: 0-3 Volts peak. Power Requirement: Two 9-volt batteries or the optional IPA-5280-1 Power Supply. Dimensions: 5.75″ H × 11″ W × 7.75″ D (14.61 × 27.94 × 19.69 cm). Net Weight: 3.25 lbs.

Fig. 8-3. Table of specifications for the Heathkit IG-5282 Audio Generator.

output is selected and matched with the appropriate auxiliary equipment. Kit building time is greatly reduced for this device because all components mount on a single circuit board. Figure 8-3 provides a table of specifications for this audio generator.

RCL BRIDGE

Generally, a test bridge is a rather expensive instrument which is not often seen on many electronic benches. The IB-5281 RCL Bridge (resistance, capacitance, inductance bridge) costs only $44.95 and features solid-state circuitry that lets you easily determine unknown values of resistance, inductance, and capacitance. Resistance is indicated in three ranges from ten ohms to ten megohms. Inductance measurements range from ten microhenrys to ten henrys, and capacitance measurements span from ten picofarads to ten microfarads. Again, a single circuit board speeds assembly. Once this is wired, the mechanical components are inserted through the front panel and the entire assembly enclosed by a molded cabinet. Figure 8-4 shows the specifications for the Heathkit RCL Bridge.

IB-5281 SPECIFICATIONS: Resistance Ranges: 10 ohms to 10 megohms, in three ranges. Inductance Ranges: 10 microhenries to 10 Henries, in three ranges. Capacitance Ranges: 10 picofarads to 10 microfarads, in three ranges. Oscillator Frequencies: 1 kHz, 10 kHz, 100 kHz. External Standard Range: 1/1 to 10/1. Power Requirement: Two 9 Vdc batteries, or optional IPA-5280-1 Power Supply. Dimensions: 5.75″ H × 11″ W × 7.75″ D (14.61 × 27.94 × 19.69 cm). Net Weight: 3.50 lbs.

Fig. 8-4. Table of specifications for the Heathkit RCL Bridge.

AUDIO SIGNAL TRACER

The Heathkit IT-5283 completes the complement of instruments included in the 5280 series. A diode-equipped rf probe is furnished with the kit and allows for the quick tracing of radio and television receiver circuits in addition to transmitter circuits as well. By using this device, component and stage failures in these types of equipment may often be quickly identified. When the probe is switched to the dc position, you are able to locate defective circuits in all types of audio components and systems. In the audible position, the signal tracer emits a tone that changes in frequency, depending upon the test-point voltage or resistance. This allows for the easy tracing of signal flow through logic circuits to isolate problems. Figure 8-5 provides the specifications for the IT-5283 Signal Tracer.

POWER SUPPLY

As I mentioned previously, a power supply is available which will simultaneously provide operating current to all five instruments. This replaces the two 9-volt dc batteries required for each. The multimeter will still require the single C cell, which was mentioned earlier. Total price for the power supply is about $25.00. This also comes as a kit which is identified as IPA-5280-1. It is highly recommended that this latter device be purchased if you intend to go with the entire 5280 series. Battery operation is fine for in-field uses, but it is a constant bother to be changing batteries on test-bench equipment. All five instruments would require a total of ten 9 Vdc batteries. Just a few battery changes would exceed the price of the power supply. It's still a good idea to keep a package or two of batteries around the shop in case it is necessary to take measurements away from the 110-volt ac source.

IT-5283 SPECIFICATIONS: Functions: Substitute speaker, Af signal tracing, Rf signal tracing, Audible volt/ohmmeter. **Speaker:** 3″ permanent magnet type. **Power Requirement:** Two 9 Vdc batteries or optional IPA-5280-1 Power Supply. **Dimensions:** 5.75″ H × 11″ W × 7.75″ D.

Fig. 8-5. Table of specifications for the Heathkit IT-5283 Signal Tracer.

ELECTRONIC WIRING

The rest of this chapter will be devoted to plans for several simple electronic test instruments which can be constructed from components that are readily available from your local electronics hobby store. But before getting into these, it is necessary to provide some general information on electronic building.

It was reported some years ago in an electronics magazine that the major cause for failure (over 90%) of electronic kits, whether commercial or home-brewed, was faulty soldering. Too many beginners and experienced technicians alike tend to view the soldering process as being so simple as to not require practice and close adherence to basic rules of building. This is simply not true. The most efficient circuit made can be rendered completely useless by improperly soldering a single contact. If you want to build electronic test instruments at home, you must know how to solder or your projects can easily end up dismal failures. Soldering is always a critical part of electronic circuit construction and must be handled carefully and with *strict* adherence to proper technique.

The soldering iron most desirable for assembling the projects that follow is the pencil type, which has a power rating of 25 to 30 watts. This provides adequate heat to get the job done but does not get so hot that the fragile solid-state components are destroyed. Soldering guns are very popular for certain types of electronic assembly, but most are rated at more than 75 watts. Soldering guns heat to temperatures which are much higher than required for the assembly of the smaller electronic projects in this chapter. Also, the soldering tips of these guns are overly large for many compact applications. Some manufacturers offer expensive soldering stations which include a pencil soldering iron, an insulated holder, and a control box which keeps the temperature of the iron constant at all times. This type of soldering equipment is shown in Fig. 8-6. While these

Fig. 8-6. A typical bench soldering iron, control box, and holder.

devices are very convenient, they are not necessary, and a simple soldering pencil from a local hobby store may be purchased for less than $10.00 (Fig. 8-7). Make certain you follow the manufacturer's instructions when preparing the tip of a soldering pencil for first use. This is usually when the tinning procedure takes place and involves heating the iron and applying a small amount of solder to the tip. When the tip is covered with a very thin coat of solder, normal soldering functions may be undertaken.

Only one type of solder is suitable for use in the construction of the projects in this chapter. This is resin core solder, which is sold by most electronic and hobby stores and is always identified as such. There is another type of solder which may be sold in hardware stores and plumbing outlets which has an acid core. The center of this solder contains an acid which is desirable for plumbing applications but will ruin electronic circuits and components. The corrosive acid-core solder usually results in cold-solder joints in electronic component leads, and the acid will gradually eat away at the delicate circuit conductors. A cold-solder joint is a connection which has not been made properly and results in high electrical resistance. High-resistance joints present most of the problems in improperly soldered electronic circuits. These bonds do not adequately conduct the flow of electrical current and can sometimes cause rectification in audio circuits. A cold-solder joint is a poor or nonexistent electrical connection. It is most often caused by simply dropping the solder onto the joint before the elements have been heated to the proper temperature. This can occur when the tip of

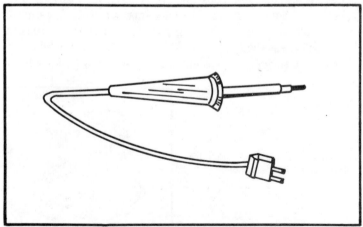

Fig. 8-7. A pencil-type soldering iron.

the soldering iron is applied to the solder rather than to the joint to be soldered.

There are several steps involved in forming proper solder connections. Each of these must be followed, in the correct order and to the letter, to arrive at a completed product which is electrically stable and dependable. The steps are as follows:

1. Make certain the elements to be soldered are clear of all foreign matter or debris. Wire conductors, for example, should be scraped clean of all insulation and wiped free of oil, tar, or grease.

2. A firm mechanical joint must be formed from the elements of the joint before soldering is attempted. This is done by tightly wrapping the conductors in such a way that no physical movement is possible between elements.

3. The soldering iron should be hot enough to allow for proper heating. It should be turned on a few minutes before soldering is attempted.

4. The soldering iron is applied to the joint, not to the solder. Once the joint has been properly formed mechanically, the soldering iron tip is placed against it to allow it to heat to the same temperature.

5. The solder is then placed against the joint, not against the soldering iron, and allowed to flow freely around the elements. When the joint is heated to about the same temperature as the soldering iron, its elements will meet the solder and allow it to flow into every part of the wrapped conductors and contacts.

6. Apply only enough solder to get the job done. Too much solder can create a cold-solder joint.

7. Once solder is flowing in the joint, remove the tip of the iron and make certain that the elements are not allowed to move. Motion at this point can cause the cooling solder to become cracked or loose in certain areas of the joint.

8. Allow about 20 seconds for the solder to cool.

9. Wiggle the protruding elements of the joint to make certain that no physical movement occurs where the solder bond has taken place.

10. Examine the appearance of the solder joint, looking for any signs of a dull surface or globular solder deposits. A proper solder joint will have a smooth, shiny appearance, while a dull, rough surface indicates a cold-solder joint.

While these steps may sound complicated upon first reading them, they will become second nature to you as you complete more and more solder connections. After only a few hours of practice, it will take only seconds to solder each joint in an electronic project. The main trick to soldering is to always apply the tip of the iron to the joint and not to the solder and to allow the solder to flow into the crevices of the joint before removing the heat. Remember to use the least amount of solder to get the job done. Cold-solder joints can result when too much solder is used because the cooling rate is uneven in the different layers of the molten lead that is applied to the joint elements. A soldering iron applied to the outside of a large blob of solder may cause only the outer portion to become molten while the inside remains relatively hard. This latter portion is the part of the solder joint which performs the electrical bonding.

A firm, mechanical joint is stressed because solder by itself does not have adequate mechanical strength to form this physical type of bonding. It serves only as an electrical bond, not as a mechanical connection. If solder is used to hold two conductors in place, for example, normal stresses will cause this connection to work loose and the solder contact to crack because proper mechanical rigidity of the joint was not originally established. Again, solder forms only an electrical joint. The elements of the joint itself must form the mechanical connection.

Even when using the low-wattage soldering pencils, speed in making the joint is often very important. Some of the solid-state devices used in the projects that follow can be damaged or destroyed when they become heated past their maximum points of endurance. If you are not experienced in proper soldering methods, you would do well to practice on a more rugged device such as a resistor, capacitor, or even upon two wire conductors wrapped together. Practice proper soldering techniques until they become second nature to you. This will increase the speed with which you're able to make the joints and is important because the longer the soldering iron is applied to the leads of a component, the hotter the component gets. A happy medium must be arrived at wherein adequate time is taken to complete a solder joint without taking so much time that the components become excessively heated.

HEAT SINKS

A heat sink is often used to aid in the further protection of heat-sensitive electronic components when soldering. This is a device which sinks or absorbs heat. A pair of needlenose pliers can serve as a very good sink when used to tightly squeeze a lead at a point near the shell of the component, as shown in Fig. 8-8. Heat will travel

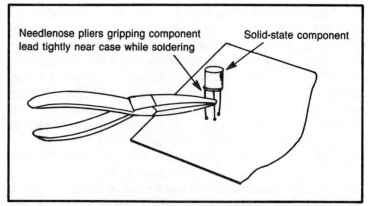

Fig. 8-8. Long-nosed pliers may be used as a heat sink during soldering.

up the lead from the point where it is being soldered, but the larger mass of the needlenose pliers will absorb most of it, which prevents a great deal of heat from reaching the case or shell of the component. Alligator clips and special heat-sink clips can also be used to form a good source of heat protection. These devices have the advantage of remaining in place after the clip contact has been made and will free the builder's hand for other parts of the soldering procedure. When applying a heat sink, make certain that it is not located too closely to the point on the lead which is being soldered. Placed too closely, the heat sink can pull heat away from the joint and create a cold-solder connection. The heat sink is best placed at a point on the lead nearest the component case. Figure 8-9 shows a heat sink attached to a transistor lead.

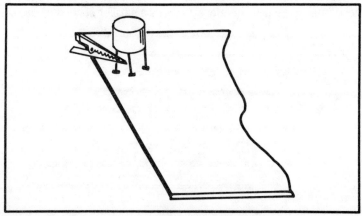

Fig. 8-9. An alligator clip makes an excellent snap-on heat sink.

Again, the proper soldering of electronic circuits is of paramount importance in electronic building. If you take shortcuts when putting together the projects in this book, you are bound to run into trouble, either when the device is first tested, or later when poor soldering connections break down. A few minutes spent in properly completing a project can save many future hours of troubleshooting, resoldering, and replacing heat-damaged components. Do not attempt to even start on a project until you know the correct methods of soldering.

When putting together any electronic project, certain skills and tools are necessary to complete the job in a manner which will provide correct and lasting operation. Many projects may be designed for installation on perforated circuit-board material, which is available at a low price from almost any hobby store. This type of construction allows for detailed assembly and provides a sturdy and compact base for all circuit components.

Figure 8-10 shows a piece of perforated circuit board. The wire leads from the components are simply pushed through a convenient hole in the circuit board. Wiring connections are made on the bottom of the board by twisting various wire leads together or connecting them by using single pieces of hookup wire and then soldering. This will also provide for a very neat appearance on the top of the board where the components are located along with a sturdiness which is not easily obtained with point-to-point wiring on contact strips.

Several ways of mounting components can be used, and the procedure followed will depend largely upon the amount of space available on the circuit board and also on the type of circuit being built. Figure 8-11 shows examples of vertical and horizontal mounting of various electronic components. Vertical mounting takes up less horizontal space and therefore requires a smaller section of circuit

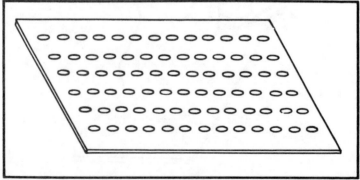

Fig. 8-10. Perforated circuit board makes an excellent building base.

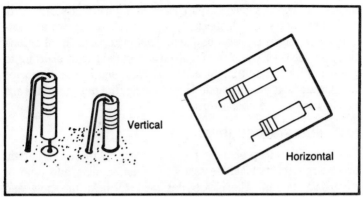

Fig. 8-11. Examples of horizontal and vertical mounting on a piece of circuit board.

board. Horizontal mounting of the components will require more horizontal space but will not require as much vertical room and provides a flat, finished circuit. This latter type of wiring is ideal for circuits which must be housed in a box or container. Vertical mounting will generally take up the least amount of space because the circuit will not be so spread out.

TOOLS

When building the projects outlined in this chapter, a normal assortment of shop tools will usually be all that is needed. Included in your complement should be needlenose pliers for bending the wire leads of the components and for wrapping them at their contact points. Wire cutters, sometimes known as diags, will also be needed to cut these leads to the correct lengths and for trimming after soldering has been completed. A pocket knife is also desirable and may be used to cut sections of perforated circuit board and for scraping insulation from the surface of painted or enamel-coated wire. A pocket knife will also be useful in clearing away blobs of solder which may accidentally drop in circuit board materials and other components. It is also good to have a wire stripper, electrical tape, and an assortment of Phillips and flathead screwdrivers of the small to miniature variety. Epoxy cement will come in handy for securing the bulkier components to the circuit board, thus preventing movement due to vibration. Other tools which are useful but not absolutely necessary include a set of nutdrivers and a desoldering tool for removal of incorrect solder connections. Alligator clips will be useful for protecting delicate components during the soldering operation. These should be purchased in a general assortment of many different sizes. A special type of insulation known as heat-shrinkable tubing can also be used. This material is fitted around the bare leads of components which are in danger of shorting to ground. This loose-fitting insulation is tightened to the conductor by applying heat from a match. Upon heating, the tubing shrinks and molds itself to the conductor.

BUILDING PROCEDURES

In addition to proper soldering techniques, there are certain procedures that should be followed when building any type of electronic circuit. These are designed to help the new and inexperienced builder become more proficient at what he is doing, to create the fewest possible problems, and to enable him to successfully complete all of the projects he attempts to build. We have all seen the results of half-completed projects. These are the ones which were started long ago and were to be completed as soon as an additional part was obtained, but which just never got finished. Half-completed projects are often subject to breakage and other types of damage because they are usually not installed in a box or a covering which provides mechanical protection.

The uncompleted project is not usually the result of lack of interest, lack of ability, or lack of skill. It is often the result of beginning a project before the builder has all of the parts necessary to complete it. This is a cardinal rule of electronic building: never begin a project until all parts, components, connectors, and the housing are on hand to complete the project. When you begin an electronic project with certain components missing, you cannot build the circuit in an orderly manner, as would be the case if all parts were on hand. The builder makes certain mental notes about parts which have been left out and which are to be replaced, and then, at a later time, completely forgets about them. A few of the components which were not on hand originally may be obtained, wired into the circuit, and then, assuming that the project is finished, power is applied. Unremembered by the neophyte builder, a component or two was omitted from the circuit, a component which the builder was supposed to have made a mental note of. Because this has been completely forgotten, the builder assumes the circuit is finished and finds that it does not work properly or at all. He now has to troubleshoot the circuit and will be more apt to look for poor connections or damaged components rather than missing parts. The circuit often winds up a total failure and is tossed into the junk box as a source for spare parts. Here is a good example of a circuit which probably would have worked perfectly if proper techniques had been followed.

While many builders will complete an electronic circuit with all the electrical components on hand, the case or box which is to house the finished product is often saved for last. There is nothing quite so fragile as an electronic circuit on a perforated board which is not protected in some manner. These boards have nasty habits of falling from work benches or of breaking when accidentally placed under heavy objects. As soon as your circuit is completed, it should be mounted in a protective case immediately after the initial testing.

A major cause of improperly wired circuits is fatigue. Experienced builders never work on circuits when they are tired, sleepy, or when their minds are on other things. If you work too long on a small circuit board, your vision will often start to blur and hands may begin to shake from being in one position too long. When you feel the least bit fatigued, stop what you're doing and take a ten to fifteen minute break until you are refreshed again. Don't set a specific day or time to have your circuit completed. When running behind schedule, you may start to rush or work past your point of adequate concentration. The result of this may be a circuit with problems such as polarity reversals of components, wiring errors, or improperly formed solder joints.

Make certain that the work bench area where you assemble your circuits has adequate lighting and ventilation for ease of construction. Arrange your seating so the normal assembly of circuits will not put you in an uncomfortable position, causing you to strain or reach in such a manner that you tire rapidly.

Most of these suggestions for good building techniques are just common sense and should be obvious to anyone. It is a good practice to have one specific area where your electronic assembly is normally done. This gets the builder accustomed to working under set conditions and makes for a more comfortable and relaxed assembly.

By following these construction suggestions, you should be satisfied each time a project is completed, both with the quality of the finished product and with its operation and dependability. You can also take pride in the fact that a great many electronic projects being built by other individuals not adhering to these techniques are going unfinished or are causing problems when completed.

IC BUILDING TECHNIQUES

All of the heating effects that create problems in building with

243

integrated circuits can be overcome by using a socket. The socket is soldered into the circuit before the device is inserted and there is no possibility of any damage occurring due to heat. Adequate time may be taken when soldering these sockets without any fear of heat damage. Proper soldering techniques are still dictated, however, as a cold-solder joint at a socket will cause just as severe a problem within the circuit and possibly more because of the added resistance created by the friction contact of the device within the socket.

If solid-state devices are to be used with sockets, it is extremely important to make certain the device leads are cleaned of any foreign materials, especially those of an oily nature. A dirty lead can form a high resistance contact within the socket and cause the same types of circuit problems that are most often brought about by cold-solder joints. The socket contacts should also be cleaned to make certain that no grit or foreign material has covered the areas that make contact with the leads. Periodic inspection of the socket is necessary, especially if the electronic circuit is used out-of-doors or in an area which is subject to dust and dirt buildup. A circuit which uses sockets is not quite as dependable as one that uses direct solder contact, so if high vibration applications are anticipated, the socket technique may not be practical.

COMPONENT MOUNTING

Integrated circuits used in most electronic instrument projects are normally of two varieties, the circular can and the DIP, which is an abbreviation for *dual in-line package*. There is a third integrated-circuit configuration which is called a *flat pack*. This last type is most often used for computer applications and is very difficult to work with in a typical home shop due to the extremely close spacing of the circuit leads.

Circular can integrated circuits look very much like transistors with many leads instead of just three. Often, a small tab will protrude horizontally from the bottom edge of the case to give some means of reference when determining the pin connections of the IC leads. The mounting of this type of integrated circuit to a circuit board is identical to the mounting of transistors, except more device leads must be contended with. This packaging is most conductive to the home builder because it allows for point-to-point wiring and does not necessarily relegate the builder to using printed circuit boards. When using the DIP integrated circuit, circuit boards are always required unless a socket is used which terminates in long wire leads instead of the normal pin connections. Wire lead extensions can be soldered directly to a DIP IC, but the chances are great that this process will damage the component because of heating effects. It is almost impossible to connect any sort of heat sink to the extremely short pins. Also, this packaging is usually accomplished with a plastic case and will melt and become disfigured under conditions of extreme heat. The use of a DIP socket will alleviate all of these problems.

The mounting of integrated circuits and other solid-state components is best accomplished by using a small piece of perforated circuit board, which is available at most radio and hobby supply stores. It is also recommended that sockets be used for any integrated circuits which are available only in DIP configurations. Figure 8-12 shows how an integrated circuit of the circular can variety can be easily mounted on this type of circuit board by inserting each of the leads through a separate hole and then soldering from beneath. This also creates an attractive finished circuit when viewed from the top side of the board. Many of the circular can packages for ICs contain a small plastic tube at the center which acts as a divider and keeps the package slightly above the circuit board. This allows for adequate ventilation on all sides of the device housing. The perforated circuit board method of construction is technically point-to-point wiring, as opposed to circuit board construction, but the perforated board acts as an excellent base or mounting platform for all components.

Figure 8-13 shows how a completed circuit might look. Note that a vertical mounting technique has been used to conserve space for the resistors involved in the structure. This is accomplished by bending the top lead of the resistor down along the side of the carbon body and clipping both leads so that the ends are even. The same is true of the mounting of small electrolytic capacitors which contain

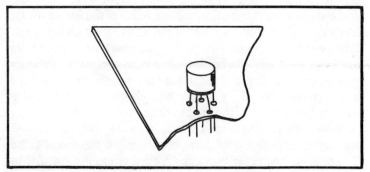

Fig. 8-12. An integrated circuit may be mounted to a circuit board by inserting its leads through the perforations.

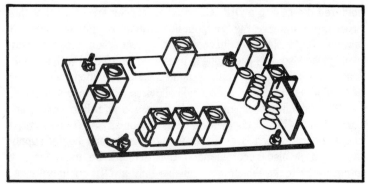

Fig. 8-13. A completed circuit board.

axial leads. These same components could just as easily have been mounted in a horizontal position (flush with the circuit board) if so desired. The vertical construction is intended to conserve horizontal or circuit-board space. All connections are made from beneath the board by twisting various leads together and then soldering them in the correct fashion.

Integrated circuits of the DIP variety will often fit in a perforated circuit board with closely spaced holes. Point-to-point wiring may be used with this type of IC if it will fit the circuit board properly, but an IC socket would be preferred. It is important to use care when installing a DIP IC in a socket. The pins of the integrated circuit are very delicate and are easily bent or even broken when forced improperly. Correct insertion procedures call for aligning all the pins on one side of the IC with the holes along one side of the socket. Notice that each pin is tapered in such a way that it suddenly becomes square at the midway point. Now, start each pin into its own socket hole, but do not seat them all the way. In other words, only the tip of each pin is started into its respective hole in the socket. Next, line up the pins on the other side of the IC into their socket holes. Make certain that the pins on the first side have not slipped from their holes while this is being done. It may be necessary to slightly bend some of the pins in order to get them to align properly. This can be accomplished with a toothpick or other small pointed device to gently force the tip of the pin into the correct slot. At this point, check all of the IC pins to make certain they are correctly inserted into each of the socket slots. Now, press firmly at the center of the IC in order to cause the remaining portion of each of the pins to snap firmly into place. A slight rocking motion when pressing the

IC may cause easier entry. Removal of the device from the socket is much less complicated and is done by simply inserting a small screwdriver under one end and gently pulling upward until the IC snaps out. Practice this procedure with a defective integrated circuit if possible, because if the pins are badly disfigured, the component may be ruined.

BUILDING SUMMARY

Although this brief discussion on building techniques dealt with circuits using solid-state devices, it should be pointed out that heat damage can occur to any electronic component. Small circuit boards generally use tiny, low-wattage components. Resistors, capacitors and coils can be easily damaged, so solder these as rapidly as possible while making sure that you don't rush to the point where cold-solder joints are formed.

Solid-state components are extremely simple devices which may be used to construct rugged, reliable systems. If strict attention is given to their selection, ratings, and mounting, the circuits that are built from them should last a lifetime. Take the necessary time to properly design your circuit and practice proper soldering techniques. Generally, you should make your circuits as simple and uncluttered as possible. This will add to the reliability and subtract from the repair time should a malfunction in the circuitry occur. Take the time to make a good solder connection, but not so long that a delicate device is damaged from the heat. After soldering is completed, allow adequate time for the devices to cool to room temperature before applying power. Many good solid-state devices have been unnecessarily destroyed by applying operating voltage and current a few seconds before they had cooled from recent soldering.

INSTRUMENT PROJECTS

In this section, several simple instrument projects are presented which the reader may choose to build from easily obtainable components. Some of these circuits are designed to act in conjunction with existing equipment, while others are completely operational alone. It should be pointed out that the solid-state designation numbers may be easily cross-referenced with components by another manufacturer. Pay strict attention to the schematic drawings and utilize the information learned in the preceding pages to make certain that each project is constructed in the proper manner.

Continuity Tester

Ohmmeters are often used as continuity testers to make sure that a particular circuit or conductor is not open or shorted. A low-resistance indication means that there is continuity, but a high resistance reading indicates a break in the conductor or circuit. Basically, an ohmmeter is simply a low-voltage source connected to a meter through the circuit under test. Figure 8-14 shows an extremely simple continuity tester which works along the same principles as the ohmmeter but uses a small light bulb instead of a meter.

This circuit is quite easy to construct and really requires only a bulb socket, an AA battery (penlight), two probes, and a few pieces of wire. This circuit need not be constructed on perforated circuit board. It's best assembled by installing the lamp socket through the side of a small aluminum box so that the globe protrudes. It may be possible to directly solder the bulb to the circuit, negating the need for the socket. Here, you must drill a suitable hole in the aluminum box and fit this opening with a rubber grommet. The base of the bulb can then be slipped through for a tight fitting. Since the grommet can easily be destroyed by the heat of the soldering iron when placed against the bulb base, it would be a good idea to solder the wires to the bulb first, then insert these conductors and the base of the bulb through the opening at the same time. The internal connections can then be made to the battery and to the probes through separate soldering actions.

You don't have to use an AA battery; any 1.5-volt cell will work. The smaller sizes, however, will keep the overall size of the entire device to a minimum. Regardless of which type of battery you use,

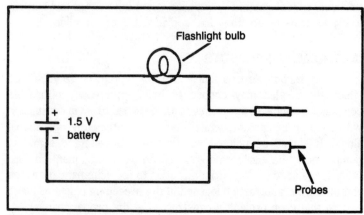

Fig. 8-14. A simple continuity tester.

it will be necessary to purchase an appropriate battery holder. These are available inexpensively at most electronic hobby stores. If you wish, two batteries may be inserted in series instead of using only one. This will provide a brighter lamp indication.

The probes are best made by connecting alligator clips to the ends of two flexible, insulated conductors. Alternately, standard ohmmeter probes may be used. Here, it will be necessary to remove the plugs and solder the conductors directly to the bulb and negative battery terminal, respectively. This circuit is totally nonpolarized, which means that you don't have to worry about making certain the positive battery connection attaches to the bulb and the negative one to the probe. If the battery is reversed, the circuit will still work.

An aluminum "bud" box is the most appropriate container for this circuit, although plastic cases may also be used. If a metal box is opted for, then the negative terminal of the battery may be connected directly to the aluminum case. The case will also serve as the connection point for the lower probe. Here, the conducting case serves as one-half of the circuit.

Once building is complete, simply short the two probes together. The lightbulb should immediately glow. If it does not, one of three things has occurred:

1. The bulb is defective.
2. The battery is weak.
3. There is a wiring error or break in one of the conductors.

This circuit is so simple that it should work the first time. Once checked out, it can be used to identify broken connections, shorted connections, etc. Never use this device on an activated circuit, one which is receiving power from another supply.

Rf Voltmeter Probe

Many electronic voltmeters are not equipped from the manufacturer to measure radio-frequency voltage. Most can be used for rf measurements, however, when the circuit shown in Fig. 8-15 is used. This is an rf voltmeter probe which is designed to attach to your electronic voltmeter in place of the standard probe leads. The mating connector at the output of this device is chosen to fit the receptacle in your present electronic voltmeter.

The circuit consists of a ceramic-disk capacitor, two half-watt carbon resistors, a germanium diode, and little else. It may be constructed on a miniature piece of perforated circuit board or may

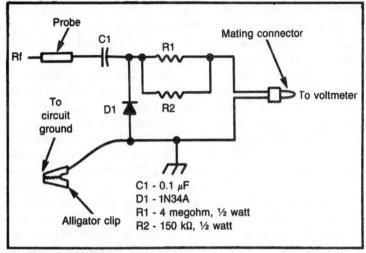

Fig. 8-15. An rf voltmeter probe.

be wired to a small terminal strip. If possible, locate an old plastic-cased probe assembly of some sort. By making the circuit as small as possible, it should be easy to insert the entire assembly into the probe body. Alternately, you may use a small case similar to the one used for the previous project. Plastic is to be preferred. A stiff piece of copper wire can be inserted through the case to serve as the probe tip. A two-conductor cable serves to link the rf probe to the voltmeter, which will read rf voltage on the dc volt scale.

Make certain the diode is inserted into the circuit as shown in the schematic drawing. A reversal here will cause the meter to read backwards. You may have some difficulty in locating a four-megohm resistor. Alternately, four one-megohm units may be wired in series with the 150 kilohm resistor paralleling this series string. Make all wiring connections as short as possible to reduce stray rf pickup.

Check over your circuit when it is completed and then insert the plug into the appropriate socket on your electronic multimeter. In the dc volt setting, you can now read rf voltages directly with a fair amount of accuracy. Actual accuracy will depend upon the internal construction and quality of components used in the electronic voltmeter and in resistor tolerance values for R_1 and R_2.

FIELD-STRENGTH METER

Figure 8-16 shows a simple rf field-strength meter which can be used to align antennas, transmitters, transceivers, etc. This is not an especially sensitive field-strength meter, so its applications will involve equipment with moderate power outputs.

The circuit should be constructed in a small aluminum box into which the small piece of perforated circuit board containing the components is inserted. Actually, only the 2.5-millihenry choke and the germanium diode need be placed on the circuit board. The five-kilohm potentiometer is mounted through the aluminum case, as is the 0 to 50 microammeter. Small links of stranded conductor connect the bottom of the choke to the positive meter terminal and the diode to the potentiometer. A stiff piece of aluminum or copper wire may be used as the sampling antenna, or the collapsible type may be salvaged from a defunct walkie-talkie. The antenna length is not critical, but it should be at least ten inches long.

Make certain the diode is wired into the circuit as shown. The same applies to the microammeter. The 2.5-millihenry choke should be located as close as possible to the antenna and the diode. Use short lengths of wiring throughout this circuit.

Once the circuit is assembled and housed in its aluminum container, simply place the sampling antenna near a moderately strong source of radio-frequency energy. Now, adjust the potentiometer for a reading on the meter. If the meter should read full-scale even with the potentiometer in the maximum resistance position, then replace the potentiometer with one of a higher value.

If the circuit should not work on the first try, recheck your wiring. Be especially watchful for a reversed diode or meter. The

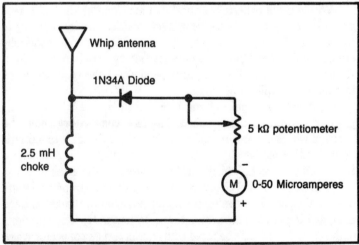

Fig. 8-16. A simple rf field-strength meter.

field-strength meter does not measure power in watts; rather, it provides relative readings of rf radiation. For instance, a full-scale meter reading at one power level will be greatly reduced at half the same power. Normally, an original reading is obtained and the potentiometer set for a mid-scale reading. Then the meter is observed while adjusting the transmitter or antenna system. If the meter reading drops, then rf power has been reduced. If it increases, the output power has risen.

NOISE GENERATOR

Noise generators are simple devices which are sometimes used in place of rf generators to align receiver circuits. Whereas the rf generator has an output at a specific frequency, the noise generator's output is very broad in nature and may span the entire high-frequency spectrum. The noise generator can be likened to an appliance motor which interferes with radio and television reception when activated.

Figure 8-17 shows the basic circuit, which is powered by a nine-volt battery. A silicon diode is used here, which can be almost any small-signal type. Don't, however, use a germanium diode. This latter type has been used for several previous projects, but will not work here. Your local hobby store will most likely have a large selection of both types of diodes.

This circuit is provided with a switch for turning the power on and off and with a control which can vary the intensity of the noise output. Coaxial cable is used as a bridge between this circuit and the receiver under test.

Wiring is not especially critical, but lead lengths should be kept as short as possible. Observe battery polarity and the polarity of the silicon diode. Component values for R_1 are not especially critical, and several junk-box potentiometers may be substituted. R_2 is chosen to match the input impedance of the receiver which in most cases will be 50 ohms.

Assemble the entire circuit (other than B_1 and S_1) on a small piece of perforated circuit board. The nine-volt battery should be supplied with a connector, and a clip is used to mount it to a plastic or aluminum case which houses the entire circuit. S_1 is a miniature SPST toggle-switch, which is also mounted to the enclosure.

The output of this circuit is connected to an appropriate length of coaxial cable. The center conductor attaches to the output lead of D_1 and the braid is connected to the bottom of R_2. The other end of the coaxial cable may be fitted with an appropriate connector which will match the input of the receiver.

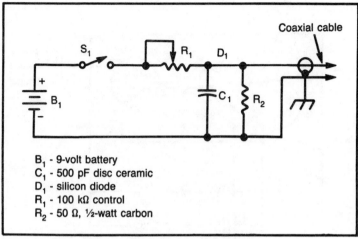

B₁ - 9-volt battery
C₁ - 500 pF disc ceramic
D₁ - silicon diode
R₁ - 100 kΩ control
R₂ - 50 Ω, ½-watt carbon

Fig. 8-17. Schematic drawing of a diode noise generator.

With the circuit connected to the receiver, activate S_1 and adjust R_1 until a frying or hissing noise is heard in the receiver speaker. The receiver may then be aligned for maximum noise output.

A bad battery, battery reversal, or reversal of D_1 can render this circuit inoperational. If problems exist, check these components as well as the overall wiring. In many applications, this simple noise generator can take the place of an expensive rf generator. The battery life of this home-brewed circuit should be very long, due to the low current drain and only occasional use.

SIGNAL INJECTOR

A signal injector or tracer is a device which is basically an oscillator whose output is connected to audio circuits to serve as an input drive. The circuit is shown in Fig. 8-18. A frequency adjust control is provided that will enable you to continuously select a wide range of audio frequencies. Two common pnp transistors are used in this circuit. The particular components specified may be replaced with any general-use pnp transistor. Npn transistors may also be used. The only modification required is the reversal of the nine-volt battery polarity. Because of the simple components required for this circuit, it may be possible for you to build the entire device from parts you already have in your electronic junk box. Other voltages may be used with this circuit, but do not exceed the nine-volt maximum.

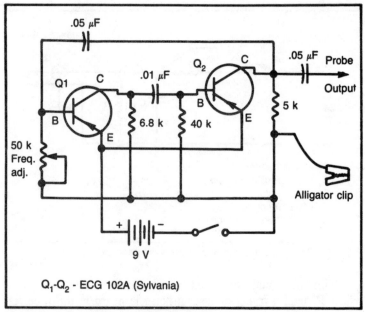

Fig. 8-18. An inexpensive signal-injection circuit.

The entire circuit is constructed on a piece of perforated circuit board, with the exception of the 50-kilohm potentiometer, which will be installed through the side of the plastic or aluminum case serving as the enclosure. The wiring is not critical, because you are dealing with audio frequencies, but as always, keep component leads to a minimum to prevent undesirable effects. The nine-volt battery should be fitted with a standard connector and clip-mounted to the inside of the enclosure. Be sure to observe battery polarity and the connections to the transistor leads. A reversal of battery polarity can easily destroy the transistors.

The output of the circuit should be connected to a salvaged voltmeter probe. The ground connection might be attached to an alligator clip lead. You may wish to use a different type of probe setup in order to suit your needs. This circuit is a little more complex than the previous one but should be easily assembled within an hour or so. Even if you have to buy all the components, the total cost should be less than $10.00.

This circuit is used to drive audio amplifiers and to trace the signal path of the injected audio frequency through the equipment under test. By placing a small earphone across the output, the actual tone may be heard.

SUMMARY

The building of electronic test instruments is a popular activity of many hobbyists and service technicians. It is beyond the scope of this book to include the thousands of circuits that are easily put together from discrete components, as these would fill many large volumes. TAB Books Inc. offers many additional books on electronic circuits which can be used directly or are readily adapted to electronic-measurement purposes.

Whether you choose to build it from scratch or from a commercial kit, it is always mandatory that you use proper building and soldering techniques throughout the entire project. Even though several hundred or even several thousand solder joints may be made in completing a piece of equipment, it takes only one bad one to render the entire device useless. A faulty solder joint is often difficult to locate after a circuit has been completed and can mean many long hours of frustrating research and reheating of all the joints before the problem is corrected.

By learning basic building and soldering techniques and adhering to their principles religiously throughout all construction projects, the home builder will eliminate over 90% of the problems that cause electronic circuits to malfunction.

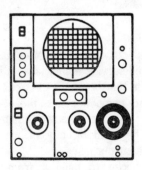

Chapter 9
Transmitter/Antenna
Measurements

The main application of the devices discussed thus far in this text is primarily that of measuring the more basic electronic units. While some of these meters may be used to measure the internal workings of a transmitter, the output power to an antenna and the antenna's measurements will often require the use of specialized measuring equipment. This chapter is devoted to radio-frequency transmitter measurements, which fall into their own special field because of their output and associated circuitry.

The measuring techniques that will be discussed apply to all types of radio-frequency transmitters, from the low-powered CB transceiver to the high-powered commercial radio transmitters. Due to the high-frequency nature of the ac output of radio transmitters, standard measuring devices such as the multimeter cannot be used to accurately measure either voltage or current. At these high frequencies, the current does not flow physically in the circuit in the same way as at lower frequencies and in direct-current applications. Skin-effect becomes more prevalent. Skin-effect is the tendency for electrical current to flow near the outside of a conductor rather than through its center or even the space throughout its entire width. Most multimeters do not offer adequate sensitivity to tiny changes in rf voltage due to the losses which occur within the meter circuitry. The meters presented in this chapter are able to perform satisfactorily under these special conditions.

THE WATTMETER

A wattmeter is used to measure electric power, which is measured in watts. Figure 9-1 shows a simplified watt meter circuit. It consists of a pair of fixed coils, known as *current coils,* and a movable coil, known as the *potential coil.* The fixed coils are made up of a few turns of comparatively large conductor. The potential coil consists of many turns of fine wire. It is mounted on a shaft carried in jeweled bearings so that it may turn inside the stationary coils. The movable coil carries a needle which moves over a suitably graduated scale. Flat coil springs hold the needle to a zero position. The current coil (stationary coil) of the wattmeter is connected in series with the circuit under test, and the potential coil (movable coil) is connected across the line.

In operation, when line current flows through the current coil of a wattmeter, a field is set up around the coil. The strength of this field is proportional to the line current and in phase with it. The potential coil of the wattmeter generally has a high-resistance resistor connected in series with it. This is for the purpose of making the potential-coil circuit of the meter as purely resistive as possible. As a result, current in the potential circuit is practically in phase with the line voltage. Therefore, when voltage is impressed on the potential circuit, current is proportional to and in phase with the line voltage.

The actuating force of a wattmeter is generated by the interaction of the field of the current coil and the field of the potential

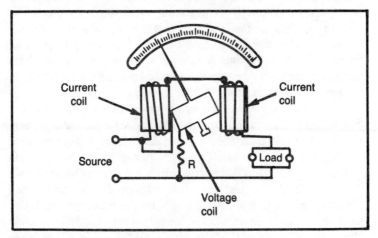

Fig. 9-1. A simplified wattmeter circuit.

coil. The force acting on the movable coil at any instant (tending to turn it) is proportional to the product of the instantaneous values of the line current and voltage.

The wattmeter consists of two circuits, either of which will be damaged if too much current is passed through them. This fact is especially important in the case of the wattmeter, because the reading of the instrument does not serve to tell the user that the coils are being overheated. If an ammeter or voltmeter is overloaded, the pointer will indicate beyond the upper limit of the instrument's scale. In the wattmeter, both the current and potential circuits may be carrying such an overload that their insulation is burning, and yet the pointer may be only part-way up the scale. This is because the position of the pointer depends upon the power factor of the circuit under test as well as upon the voltage and current. Thus, a low power factor circuit will give a very low reading on the wattmeter even when the current and potential circuits are loaded to the maximum safe limit. This safe rating is generally given on the face of the instrument. A wattmeter is always specifically rated in volts and amperes, not in watts.

We need some way to measure the output power of a radio transmitter in watts. The wattmeter can be utilized to do just this. The meter is normally coupled to the output of a transmitter. It is connected in series with the output transmission line and samples a very minute amount of the output power. This alternating-current source is fed directly to a dc milliammeter which is calibrated against a known power source in watts, kilowatts, or milliwatts, depending on the range covered by the particular wattmeter being used. The wattmeter, then, works in a manner which is similar to any other meter in that it samples a small portion of the component to be measured without having any major effect on the circuit. This instrument eliminates many of the calculations needed for determining accurate readings of power output in watts.

Figure 9-2 shows the proper placement of a wattmeter in the line, which continues on to the transmitting antenna. Actually, only the sampling unit is installed in the rf line. Figure 9-3 shows what this might look like. Basically, it is a tiny pickup antenna which drains a small portion of the rf output power from the transmitter and sends it down a cable to the rectifier and metering circuit. This circuit then converts the power into driving pulses for the milliammeter. If this sampling unit picks up too much of the output power, it will adversely affect the operation of the transmitter and the readings which are obtained will be inaccurate.

258

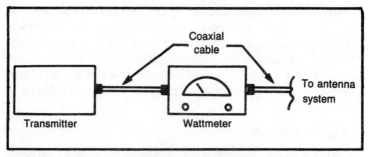

Fig. 9-2. The wattmeter is placed in the transmission line near the transmitter.

Wattmeters are available from many different manufacturers. Generally speaking, though, the higher priced units will offer greater accuracy and less interference with the rf line. Most of these quality instruments are designed for use with a specific type of radio frequency transmission line, usually those that exhibit an impedance of 50 to 75 ohms. If a wattmeter is connected into a line with different impedance specifications, the indicated power values will not be as accurate. Larger deviations in accuracy will occur in direct proportion to the amount of the difference in the line impedance from what is specified for the wattmeter being used.

Many wattmeters also serve as nonradiating antennas. These are known as *dummy loads* or *dummy antennas* and simply dissipate the output power from the transmitter similar to heat through a resistor. This type of meter is extremely accurate due to the fact

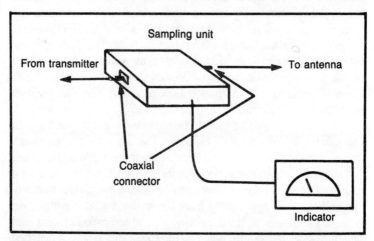

Fig. 9-3. Many wattmeters consist of a sampling unit which is placed in the line. A remote indicator is connected to the sampling unit for obtaining readings.

that the dummy load establishes a specific impedance of transmitter load with which the wattmeter is specifically designed to operate. Rather than connecting the transmitter to a radiating antenna, it is connected to the dummy-load/wattmeter combination. This instrument will provide the most accurate readings of any of the power-measuring devices used in radio-frequency transmitters.

THE SWR METER

When aligning antennas and tuning them for proper operation at specific frequencies, it is usually necessary to obtain an indication of how well the transmitter is matched to the transmission line and antenna. This calls for the use of the SWR meter, or *standing wave ratio meter*. Standing wave ratio is an electronic term which describes the ratio of the value of current or voltage at a loop to the value or voltage at a node. The SWR depends upon the ratio of the resistance of the load connected to the output of the line to the characteristic impedance of the line itself. Thus, the standing wave ratio provides a direct indication of the degree of mismatch along a transmission line. In simpler terms, it is a measurement of the amount of power which is fed to the antenna and reflected back from the antenna to the transmitter.

When an antenna is not tuned properly, the reflected power will be larger. A perfect match of the transmitter to the transmission line and antenna will result in an SWR reading of 1:1. Only the top portion of this ratio will be shown on most SWR meters, so a reading of 1, which is actually the lowest reading available on the scale and corresponds to a 0 on most meters, will indicate an SWR of 1:1.

Figure 9-4 shows a typical SWR meter. Notice that this particular instrument contains a sensitivity control as well as a two-position switch. One position is marked FOR and the other is marked REF. These abbreviations are for forward and reflected. The REF position indicates the SWR reading.

SWR meters are designed to be operated in coaxial cables with a specific impedance, normally 50 ohms. This is a standard impedance for the coaxial cable which is used in modern transmitting applications. For the most accurate measurements, the SWR meter should be installed as close to the antennas as possible, although this is rarely convenient. Most applications have the meter installed quite close to the transmitter. This will provide a good indication of how well the entire antenna system, including the transmission line, is matched to the transmitter. For most applications, this is a very adequate

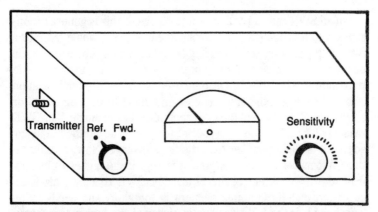

Fig. 9-4. Most SWR meters contain sensitivity controls and switches which allow for the reading of forward and reflected power.

measurement which can be used to adjust the antenna length properly.

The SWR meter must be connected into the line in the proper manner in order to obtain proper readings, because the circuitry which comprises this instrument consists of an input and an output. If these two terminals are reversed, readings will not be obtained. To operate the meter, apply a small amount of rf power to the antenna with the meter in the line and with its two-position switch in the forward (FOR) position and the sensitivity control turned fully counterclockwise. Once power has been applied from the transmitter, adjust the sensitivity control for a full-scale reading. The meter is aligned by adjusting the sensitivity control until the needle indicator is matched to the reference point on the meter scale (REF). This is not to be confused with the previously mentioned REF indicator on the two-position switch. The REF referred to here will be located on the meter scale. Once the meter has been adjusted so that its needle is calibrated to the reference point, turn the two-position switch to the REF point to obtain the standing-wave-ratio reading. The lower the needle drops to zero, which is indicated as a reading of one on the scale, the better is the match of the transmitter to the antenna.

As mentioned earlier, the SWR meter finds most of its usage in these applications in the tuning of the antenna. The tuning may be accomplished by adjusting a variable capacitor at the antenna. However, it may be necessary to actually lengthen or shorten the antenna elements. This is done by cutting or by soldering on an extra length of wire.

If an SWR reading of 2 or less is obtained, this is generally considered to be adequate. As explained earlier, this would indicate a ratio of 2:1 or less. If the reading is higher than 2, however, it will be necessary to adjust the antenna for a better match to the transmitter. The SWR can also be used to obtain a rough indication of whether or not the antenna elements need to be lengthened or shortened, if this type of tuning is applicable. For example, if an SWR reading of 3:1 is obtained at a specific frequency, adjust the frequency so that it is slightly higher than the first readings. Now, recalibrate the SWR meter and take another reading. If the higher frequency from the transmitter results in a lower SWR reading, this is an indication that the antenna element is too short and needs to be lengthened. If the SWR reading is higher at the higher frequency, then the antenna is too long and needs to be shortened. When changing the frequency, keep it in the general range that was used for the first measurement. For example, if the antenna is to be tuned for 7.5 MHz and the reading is high at this point, the frequency should be increased to about 7.6 MHz for the second reading.

Most SWR meters may also serve as a relative power indicator within the transmission line. By throwing the switch into the forward position and adjusting the sensitivity control for a midscale reading when power is supplied from the transmitter, an increase or decrease in output power will be indicated by the needle moving past or below the midscale reading. This scale is not calibrated in watts and serves only as a relative indication of power output. This output is relative to the power that was necessary to drive the meter to a half-scale reading. The sensitivity control will also determine how much power it takes to arrive at this reading. It should be remembered that each time an SWR measurement is taken, it will probably be necessary to recalibrate the meter. This takes only a moment or so and assures an accurate reading each time the instrument is used. The alignment procedure calls for a signal from the transmitter of either a continuous wave, such as that used for code transmission, or any other type of signal which is steady in nature. This continuous wave is what is transmitted by an AM transmitter when the mike is keyed but no audio is applied to it. A single-sideband transmitter does not generate a carrier, and rf energy is present only when audio is present at the microphone input. Most single-sideband transmitters, however, have the ability to generate a solid carrier for test purposes or can transmit an AM signal. It is this stable energy which the SWR meter acts on and which gives the accurate SWR singular measurements.

When using the SWR meter, always make sure that it is connected in line with an impedance which is equal to the value with which it was intended to operate. Most meters are designed to operate in a 50-ohm line, as mentioned earlier, but will give relatively accurate readings in a 75-ohm transmission line. Some meters are adjustable and can be set up to operate into both of these values of line impedance. Most SWR meters are also frequency sensitive. An SWR meter which is designed to operate only in the high-frequency region of the radio spectrum will not be as accurate in the very high frequency region, although most will provide a ballpark indication of standing wave ratio on the line. Further meter accuracy will be obtained if the meter is placed at the transmitter and the cable running from the meter to the antenna is exactly one-half wave long at the operating frequency. This is usually not necessary for most applications, in which a slight inaccuracy can be easily tolerated.

Make certain when using the SWR meter that the sensitivity control is always adjusted so that the meter is calibrated in the forward position before attempting to take a measurement. Always return the sensitivity control to zero when further measurements are not to be taken. This could prevent meter burnout should it be used again with a higher powered transmitter. The increased power output with the sensitivity control in too high a position could damage the meter.

The SWR meter shown in Fig. 9-4 is typical of the design manufactured today. However, many different styles are available, with some even having two meters. One is purely for relative power readings, while the other is for the conventional SWR readings. In this type of instrument, the two-position switch is deleted and the sensitivity control is used to align one meter to the calibrated or reference position while the SWR is read simultaneously on the other meter. Many of these meters are designed to be left in the line at all times, and some may even be calibrated in watts in order to provide a rough indication of power output. This indication will be accurate only when the line impedance and the antenna impedance match that of the transmitter and SWR meter.

THE FIELD-STRENGTH METER

Field strength, or *field intensity*, as it is sometimes called, is the effective value of the electric-field intensity produced at a point by the radio waves from a particular station. Unless otherwise specified, the measurement is assumed to be in the direction of maximum field intensity. Seldom are the actual operational characteristics of an antenna exactly the same as those determined on the basis of theoretical considerations. In order to determine these differences, various measurements must be made after an antenna is installed and while it is being test operated. Often, on the basis of these measurements, changes are made in the design or installation of the antenna to improve the radiation pattern.

It is very important to know the direction and intensity of the power being radiated from an antenna. To determine these values, measurements of the field strength are made at various distances and directions from and around the antenna.

Figure 9-5 shows a typical field-strength meter. While some of these instruments are calibrated in millivolts and microvolts, most

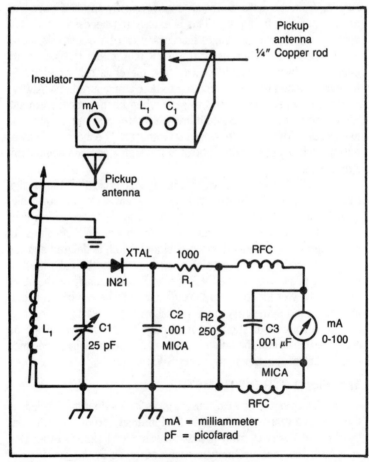

Fig. 9-5. A typical field-strength meter and its circuitry.

use a small dc microammeter with no set scale division to give an indication of what level of power the antenna is at when calibrated for a midscale reading at a specific distance from the antenna. The field-strength meter picks up the transmitted radio frequency on a small antenna in much the same manner as a conventional radio in the home does from a radio station. The radio energy is rectified, which changes the dc current. It is then fed to the microammeter, which gives an indication of the amount of current. This, in turn, is directly related to the strength of the radio-frequency signal.

By moving to different locations with respect to the antenna's position, the relative strength of the radio-frequency signal can be noted. A graph can be drawn plotting the various strengths obtained from the meter readings. Figure 9-6 shows the antenna radiation pattern which might be obtained from a vertical antenna when readings are taken on a field-strength meter which is spaced an even distance from the antenna in different directions. A vertical antenna under ideal operating conditions of flat terrain and no interfering objects will produce a signal of the same strength in all compass directions, as long as the same physical distance is maintained. In other words, the antenna is used as the center of a circle and readings are taken at various points around the circle at equal distances from the center. All readings obtained should be the same at all these different points. If, however, the circle were made smaller, the effect would be to increase the signal strength at all points, but the readings

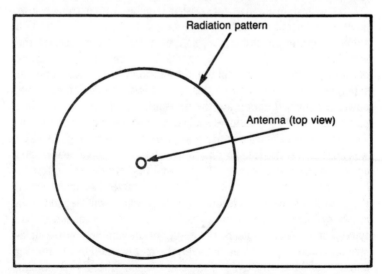

Fig. 9-6. The radiation pattern of a perfectly functioning vertical antenna.

would still be the same. Remember that we are discussing a vertical antenna operating under ideal circumstances. Practically speaking, an antenna will have some interfering objects within its radiation path, and the field-strength meter can be used to determine the effect that these obstructions have on the antenna. Also, there are some types of antennas which do not radiate power equally in all directions. A field-strength meter used with an antenna such as this will provide an indication of proper directionality by measuring the output in all directions to insure that the antenna is radiating stronger in the direction that it was intended to and weaker in the direction where the antenna is not designed to transmit.

A field-strength meter may also be used to tune an antenna by adjusting its element for maximum signal strength at a remote point. The tuning of an antenna for lower SWR readings is not always entirely accurate, while a field-strength meter will always indicate the highest amount of output power obtained with a certain antenna element length. The output of a radio-frequency transmitter may also be tuned and adjusted according to readings obtained on a field-strength meter. I have already mentioned that the best meter or measuring instrument is one which has the least effect on the circuit under test. The field-strength meter suits these purposes quite well, since it is placed anywhere from several feet to several thousand feet away from the antenna to be tested. For all practical purposes, it has absolutely no effect whatsoever on the antenna circuit.

Some field-strength meters designed for use with extremely low-powered transmitting circuits have an extra stage of amplification within their internal circuitry. This type of instrument receives the very low rf input which is then rectified and amplified many times in order to provide a readable meter indication. An external power supply will be necessary with this type of instrument, but no actual power is required to operate the remainder of the circuit, which is powered entirely by pickup of radio frequency on the whip antenna. This instrument is similar in many ways to the old crystal radio which was an rf power device. As a matter of fact, if an earphone is placed across the meter contacts, the transmitting station can often be heard. The distance at which a field-strength meter can operate from the source of the radio-frequency transmission will depend on a number of facts, such as the power output of the transmitter, the type of antenna being used, the length of the antenna on the field-strength meter, the terrain between the meter and the transmitting antenna, and the sensitivity of the metering circuits. For the tuning of mobile antennas such as those used by ham radio or CB operators, it is common practice to place the field-strength meter on the opposite end of the automobile from the location of the transmitting antenna.

THE RF AMMETER

The rf ammeter is an instrument which is inserted within the rf line in a series connection in the same manner that a dc ammeter is inserted in the dc line. When coaxial cable is used at the output of the transmitter, the ammeter must be placed in the hot lead or center conductor. The ac ammeter is used to determine power output by measuring the amount of rf current traveling to the antenna or dummy load from the transmitter. It is necessary to know the impedance of the antenna in order to be able to accurately compute power output.

Some calculations are necessary to arrive at the power output to the antenna. For better accuracy, the meter should be installed as close to the antenna as possible to determine how much power is actually reaching the antenna. If you only need to obtain the output power of the transmitter, then the meter is placed as close to the transmitter as possible, but still within the rf line. It may even be located within the transmitter chassis. The reason for the two placement points is because a certain amount of power will be lost in the rf line. Thus, if the antenna power input is to be measured, the meter would be closest to the antenna. When you need to measure the power output of the transmitter, the meter is nearest the transmitter. Figure 9-6 shows the various placement points of the rf ammeter in the rf line.

Rf ammeters are available in many different ranges. Therefore, as with other measuring devices, it will be necessary to know the approximate range of current within the rf line before a meter is chosen. The meter is often provided with a shorting device across its two terminals which, when connected, completely shorts the meter circuit out. This allows the rf energy to travel around the meter instead of through it. The shorting device usually takes the form of an insulated switch or bar. When measurements are to be taken, the short is removed and current flows through the meter.

In order to calculate rf output power, you must apply Ohm's law. According to Ohm's law, $P = I^2R$, where P (power) is equal to the value of line current squared (I^2) times the antenna impedance (R). Assuming an antenna impedance of 50 ohms and a line current reading of 4 amperes of rf current, then power in watts is equal to $4^2 \times 50$ or 16×50 or 800 watts. This formula is not all that difficult to use, especially if a small calculator is available. It will always provide quite accurate readings, as long as the resistance is expressed in ohms and the current is expressed in amperes. The only inaccuracy that may be present is usually because of dirty contacts which heat up due to the rf current passing through the meter and the line.

This method of measuring power is still used by most commercial radio stations today, but most other applications will use the in-line wattmeter discussed earlier. This instrument provides readings which are quite accurate when the antenna is being matched to the impedance of the device.

The rf ammeter will cause a slight drop in current when it is inserted into the line. This can be detected in relatively high-powered operations where line currents often exceed five amperes. It should also be remembered that the rf ammeter, when placed in the rf line, is a natural part of the transmission-line circuit. Should this meter develop a defect that causes its circuits to open up, there will be an open circuit in the transmission line and radio energy will not be delivered to the antenna. When a shorting switch is used to remove the meter from the circuit (during times when measurements are not needed), the problem of a defective meter can be remedied by simply throwing the switch to the shorting position until a new meter can be inserted.

FREQUENCY COUNTERS

While a frequency counter has applications in measuring audio frequencies, it can also be used to measure radio frequencies as well. The more expensive frequency counters are usually designed for rf work, and rather inexpensive models are available in frequency ranges of over 150 megahertz. The frequency counter is coupled to the output of a transmitter in much the same way as the wattmeter, thus sampling a very small portion of the output power. The frequency counter, however, receives unrectified rf energy at its output, while the wattmeter receives rectified radio energy from its in-line probe.

The frequency counter is designed to be used only with steady-state rf energy, such as a continuous carrier generated by a code transmitter or by an AM transmitter with no modulation applied. The counter will indicate the carrier frequency of the signal. Should audio be applied to this carrier, as in the case of an actual voice transmission by AM radio, the rf energy output becomes too complex to be read by the counter and the indication will be very erratic. Modulating a carrier can cause damage to some types of rf frequency counters. A single-sideband signal is present only when audio is fed to the microphone input. The frequency of a single-sideband signal can be read by broadcasting a single tone of known frequency through the microphone input. A single-sideband output, when triggered by a 1000-hertz signal, will be one kilohertz or 1000 hertz higher or lower than the actual frequency would be if a carrier were present. This will depend on whether or not upper or lower sideband transmission is being used. Most single-sideband transmitters have the capability of carrier insertion, and this is a more reliable means of taking frequency measurements.

When used for radio frequency measurements, a frequency counter should be well-stabilized before accurate measurements are attempted. Many of these instruments use a crystal-controlled reference frequency, which is determined by a crystal oscillator within the frequency counter circuitry. A half-hour or so of actual operation of the counter is required before the temperature of the circuitry has stabilized to the point where it will be possible to obtain accurate frequency measurements. Many modern frequency counters available on today's market keep the oscillator under power at all times, even when the on-off switch is in the off position. This switch merely deactivates all of the other circuitry within the counter but keeps the oscillator (or clock as it is sometimes called) in operation as long as the line cord is connected to an outlet. These types of counters will not require the half-hour warming-up period before accurate measurements can be taken. Be certain that adequate feeds of radio-frequency energy are made available to the counter at all frequencies. Radio energy of too low a value will cause erratic and unreliable readings. By the same token, be sure that the power input of the counter is not so excessive that it overdrives the unit and causes damage. Only a very small amount of rf energy from a transmitter is required.

The method for taking frequency measurements of rf output is basically the same for all frequency counters. The power output of most transmitters is sufficient. However, it may be necessary at times to read the frequencies of the local oscillator and other low-power circuits which make up the transmitter stages. These circuits do not usually provide enough power output to drive the frequency counter. Probes are available which contain wide-band amplifiers. These will take the very low power, amplify it, and feed it to the frequency counter at a level which the instrument's circuits can use. Most of these probes are designated with upper frequency limits

269

and may be purchased in ranges which fulfill the test requirements for individual applications. Frequency counters themselves have limited ranges which start at a few cycles on the low end and range upward to several thousand megahertz for the more expensive instruments. When a frequency to be measured is higher than the operating range of the counter being used, circuit losses within the instrument become great and accuracy will be questionable for frequencies just outside the normal operating range. Measurements will be completely wrong for frequencies in the higher spectrums.

For frequency counters with limited range, devices are available which multiply the upper frequency limit. Actually, the process involves division rather than multiplication. For example, the frequency counter may take a 200 megacycle input, divide it by 2, and retransmit it at 100 megacycles at its output. In this way, a 200-megahertz frequency may be read on a 100 megahertz counter with accurate results. Some extrapolation is necessary, but this is easily done without resorting to a calculator.

Frequency counters which operate in the high-frequency regions usually have two positions of measurement. One will be read in megahertz, while the other will read in kilohertz, with the basic unit of megahertz understood in the second position. This means that to read 7.123456789 megahertz on a meter which has a five-digit readout, the counter select switch would be set in the megahertz position. The readout would indicate 7.1234 megahertz. The switch is then set to the kilohertz position.

The second reading displayed would be 56789, which is the last five digits of the reading. This is what is meant by megahertz readings being understood.

The owner's manual which comes with each frequency counter will be a bit more explicit about how these instruments operate than the general discussion here. Each counter has its own special features as well as disadvantages, and each may operate in a slightly different manner. It should be remembered that accuracy will be affected when using external devices with a frequency counter. When using a frequency divider, for example, the stability or instability of the circuitry used for this divider will directly affect the accuracy of the reading. The direct method of frequency measurement will be the most accurate. This is when the output from the frequency course to be measured is adequate to directly drive the frequency counter. It also assumes that the frequency to be read is within the normal operating range of the instrument.

Most frequency counters will measure frequencies starting just above direct-current and extending well up over a thousand megahertz. The less-expensive models on the market sell for under $100 and will measure frequencies up to about 40 megahertz, which extends into the very high frequency region of the radio spectrum. Counters to measure higher frequencies than this require more complicated and critical circuitry and are more expensive. These instruments normally include an input which samples a very small portion of the alternating-current power source. This is done without actually interfering with the function of the circuit being measured. For example, a small portion of the radio-frequency output of a transmitter might be fed to the input of a frequency counter and the frequency measured at all times. The frequency counter is then connected in parallel with the circuit under test. Figure 9-7 shows a frequency counter of the digital readout variety. A probe is often used for ac circuits which contain power outputs in the very low range.

Frequency counters often use highly stable quartz crystals to establish an internal frequency rate which is of known value and tolerance. The incoming frequency is combined and compared with this known frequency and the difference is displayed on the digital readout as the actual frequency. Other types of frequency counters use the frequency of the ac line from which they are powered as a time base or reference.

THE IMPEDANCE BRIDGE

Though normally used only in the more demanding measurements, the impedance bridge has a definite use in modern measurement techniques. *Impedance* is the total opposition offered by a circuit to the flow of alternating current of a particular frequency. Impedance is a combination of ohmic resistance and of capacitive

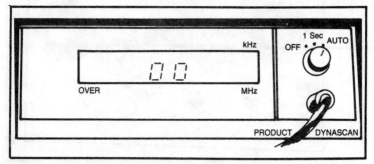

Fig. 9-7. A bench frequency counter with digital readout.

271

and inductive reactance. Impedance is measured in ohms. Although this is the same unit of common resistance, the measurement of impedance is different from that of pure resistance.

While a simple ohmmeter will measure dc resistance (the resistance of a circuit to the flow of direct current), it will not measure impedance. To measure circuits for specific values of impedance, it is necessary to have a test instrument which can produce a flow of alternating current within the circuit under test. The impedance bridge, which is shown in Fig. 9-8, does just this. While not all impedance bridges contain their own alternating-current source, all of them are designed to operate with an external generator as a source of alternating current.

An impedance bridge provides extremely accurate measurements of resistance, capacitance, and inductance. It will also measure the dissipation factor of capacitors and the storage factor of inductors. While actual use of various types of bridges varies, most of them require that the operator monitor the alternating current, which is in the form of an audio signal on a meter or in a set of headphones, while adjusting the various controls until a complete null is obtained. This null is the cancelling of the audio signal. When the signal is no longer heard, the calibrated controls are consulted and the various readings obtained. Impedance bridges can be rather demanding as far as a delicate touch is concerned in tuning for a null. It may be necessary to adjust the various controls repeatedly, as one will usually have some effect on the others.

The impedance bridge can be used to accurately measure the impedance of an antenna while being fed rf energy. This instrument

Fig. 9-8. The Heathkit Impedance Bridge.

is used by many commercial radio stations and in other situations where the close measurement of antenna impedance is required in order to meet Federal specifications. The impedance bridge is connected to an antenna directly and is fed by the transmission line from the transmitter. By connecting the instrument, adjusting the dials for a minimum reading, and then noting the calibrations on each dial, the impedance of an antenna can be read for the frequency of operation while noting the J-factor. The *J-factor* is the reading of conductive or capacitive reactance.

The impedance bridge may also be used as an SWR meter and may be inserted in the line via the transmitter to give an indication of the input impedance the transmitter "sees" at its output. This instrument is a somewhat specialized device and is normally used only in applications which demand critical measurements. It is dis-·cussed here so that you will be familiar with its operation and significance with regard to the rf measurement field.

SUMMARY

The measurement of radiation and power output in transmitter/antenna systems is not very complex in most applications. In comparison with other types of measuring and testing, relatively few instruments are needed for this purpose, and even then, they are generally simple devices which cost very little. Most people with some experience in citizens' band or amateur radio have had the experience of performing some type of transmitter/antenna measurements. Perhaps there are a few instruments in this chapter which you have not yet taken advantage of. These are just as easy to operate as the SWR meter. Many SWR meters also include circuits which internally convert them to field-strength meters at the flip of a switch. The field-strength meter can be an even better determinant of proper antenna matching and power output than the SWR meter. In the field-strength mode, the measurement of radiated power is a slightly less complex operation than the measurement of the standing wave ratio (SWR).

For more exacting requirements and measurements, an impedance bridge may be necessary. This requires a bit more practice for proper operation than the other instruments mentioned above, but even this device can be mastered in an hour or so. Many hobbyists and home technicians shy away from comprehensive power and antenna measurements because of their lack of knowledge about what is involved in the overall process. Once proper power/antenna measurement principles are learned, this fear will quickly disappear.

Chapter 10

Distortion Measurements

In the production of audio output, a certain portion of the signal (most of it) makes up the main signal. However, there will always be alternate signals produced which are usually quite complex in nature. This is because they are made up of many different tones and sounds. This action is a natural by-product of the generation of the pure sine-wave tone, or frequency. The frequencies that are generated in addition to the one which is desired are known as *distortion*.

Distortion is the unwanted part of an audio signal. It can be caused by a number of factors. Some of these include operating the electronic circuit at a level above the normal operating curve, overdriving an electronic circuit, or possibly a faulty circuit component. Of course, to a certain extent some distortion always exists in any signal. Because of this, it may sometimes be necessary to measure the distortion content. Although distortion is often thought of as being just static and noise, it is composed of pure sine-wave signals which are produced as a by-product of generating a specific signal. Distortion is also a change in the desired waveform of a signal.

The measurement of distortion is expressed as a percentage. This percentage indicates how much of the signal which is transmitted is actually distortion. In other words, if a signal is 75% pure and the other 25% is composed of distortion, then a measurement of distortion in this particular situation would be 25%. This is used as an example, but a distortion level of 25% is quite high and could easily be heard by the human ear. In audio signals, this would be detected as a garbling or a fuzzy effect on the signal which is heard.

Generally speaking, most distortion measurements are used to detect much smaller percentages of unwanted signals than in this example. These are usually on the order of anywhere from 10% to less than 1%. Commercial broadcast stations are often required by the Federal Communications Commission (FCC) to take distortion measurements. Regulations set forth by the FCC specify that the signal which is broadcast by commercial stations must have an audio distortion content of less than a certain figure. In order to meet these legal requirements, these transmitting facilities are usually required to take distortion measurements and then report them to the FCC at specified intervals.

Manufacturers of commercially available stereo components, public address systems, etc., are also required to provide specifications on each piece of equipment sold. These specifications usually include a figure (expressed as a percentage) which measures distortion. The amount of distortion content in a piece of equipment, especially in a sound system, is of prime importance when choosing from a number of possible systems.

Another use for the distortion-measurement test that may be of particular interest to the home experimenter is in the testing of projects once they are completed in order to make sure they meet the required specifications. It may also be necessary to take a distortion measurement in order to find the cause of a malfunction.

Although the distortion-measurement test is not the one used all the time, it is certainly an important and helpful aid to both the home experimenter and the technician as well. This chapter will outline the operation of the distortion meter and the actual procedures involved in taking distortion measurements.

THE DISTORTION METER

The distortion meter is an instrument, which, as its name implies, is used to measure distortion content in audio circuits. It is important to know that this is a frequency-sensitive instrument and therefore must be set to respond to the range of frequencies which are fed to its input. While the instrument has both an input and output, most distortion meters can be operated without any connections to the output terminal. The output is provided as a means for measuring distortion while the unit is in series with the audio system. This enables normal operation to continue at the same time measurements are being taken.

Once the frequency range control has been set to cover the range of frequencies which will be fed into it, the audio signal is connected to the input. There are several percentage controls provided on a basic distortion meter. For the initial testing of the audio circuit, this control is set in the reference position. When the audio signal is fed into the unit, a sensitivity control is adjusted so that the audio signal will produce a full-scale reading on the meter face. Once this is obtained, the percentage switch is thrown to the top percentage position. The meter indicator is then nulled by adjusting the amplitude and frequency-adjust controls for a minimum reading. The meter is nulled when any adjustment of the two controls will not produce a lower reading.

Distortion meters generally have four controls. Two controls provide for coarse adjustment, while the other two serve to provide a means of fine adjustment. The coarse adjust controls are adequate for the higher percentage scales, but the fine adjust controls will be necessary for the low percentages, where only a slight movement of these controls will send the needle indicator off scale.

Once a null has been accomplished, the range percentage switch is decreased a step at a time for more sensitive readings. Each time the switch is flipped to a lower range, the null adjustment controls are turned to the low reading. When it is not possible to decrease the percentage scale without setting the needle off the meter scale, this indicates that the proper distortion measurement has been arrived at. The measurement is read on the scale which is normally calibrated from 0 to 100. It is read as 0 to 100 all the way down to 0 to 1 by adding or subtracting zeros from what is printed on the face of the meter. When the range switch is in the 1% position, then, the far-right reading of 100 on the scale is equivalent to a reading of 1. These graduations are measured in percentages.

As the percentage range switch is lowered to the more sensitive position, a very delicate adjustment of the fine-tuning controls will be necessary. The distortion meter is extremely sensitive at this time. Even slight changes in the adjustment will produce extreme changes in the meter indicator.

For distortion measurements, the input audio signal to the distortion meter must be a fairly constant amplitude or volume. Its frequency must also be stable as well. Any changes in these values will result in erratic readings. In order to take accurate measurements, it is necessary to allow the distortion meter an adequate amount of time to reach its proper operating temperature.

Before any signal is fed to the input of the meter, the control must be set at zero.

THE SIGNAL GENERATOR

When taking distortion measurements, it is often necessary to use a signal generator in conjunction with the distortion meter. A signal generator generates audio signals. These signals or tones of steady frequency are needed to obtain accurate test readings. The audio oscillator will generate tones as low as 20 Hz and as high as 10,000 Hz or more. The output from this signal generator can be increased or decreased, and many of these units contain a built-in meter which serves to measure the output in decibels (db). The frequency of the meter is adjustable and is calibrated in hertz (Hz). In order to produce a 1 kHz signal, the control is simply set to the 1 kHz position and the output adjusted for the desired level.

The more sophisticated signal generators, or *audio generators* as they are sometimes called, boast a very pure output waveform. In other words, there will be very little distortion generated. Since the purpose of the signal generator is to perform distortion tests, it will be necessary to determine what distortion value is present in its output signal.

A simple and basic arrangement for measuring distortion involves coupling the output of the signal generator to the input of the distortion meter. The generator's frequency control is then set to a frequency which is within the range of the anticipated distortion test and a standard distortion test can then be run using the distortion meter. The distortion level will usually be less than 1%, but this figure must be known so that it can be subtracted from the total distortion obtained when running a distortion test on another device. The figure arrived at, let's say 1% in our example, indicates the amount of distortion present in the signal generator. This must be either added or subtracted from the figure arrived at when testing another piece of equipment.

Another typical arrangement for measuring the distortion content of an audio amplifier uses both the signal generator and the distortion meter. The output from the signal generator must be matched to the input requirements of the audio amplifier for accurate test results. If this is not done, a mismatch at this point will create distortion. The output from the audio amplifier is then connected to the distortion meter. Assuming the audio generator has a distortion level of .5%, the test begins. The audio generator is set for a frequency which can be chosen at random, but for this test, it will be 2000 hertz.

The frequency-range control on the distortion meter is set to include a 2000-hertz signal. With the percentage range switch in the reference position, a signal is fed from the signal generator to the audio amplifier which, in turn, feeds a signal directly to the distortion meter. The sensitivity control on the distortion meter is adjusted, along with the output volume of the audio amplifier (if necessary) to produce a full-scale reading on the distortion meter. When this is accomplished, the percentage range switch is dropped to the percentage position. The course amplitude and frequency controls are then adjusted for a minimum reading, and the fine adjust control is also tweaked to make certain the distortion meter indicates the minimum. When the tweaking procedure is repeated, the percentage control is dropped to a more sensitive position. When the meter will no longer indicate a lower reading, check to see which sensitivity position is indicated and take a reading on the meter. For example, if the meter indicates a 10% range and the needle indicates an 80 on the scale of 0 to 100, does this mean the amplifier has a distortion percentage of 8%? No, it doesn't, because this figure is a measurement of the total distortion present in the audio-signal generator and the audio amplifier combined. Remember that the distortion measurement of the signal generator, which has already been taken, is .5%, or ½ of 1%. In order to determine the total distortion which has been added by the introduction of the audio amplifier, this .5% figure must be subtracted from the total distortion measurement. Therefore, the distortion present in the audio amplifier is 7.5%. This is the correct figure rather than the 8%, because what is needed is the amount of audio distortion which is created by amplifying a completely pure audio signal by the audio amplifier under test. While a completely pure source of signal generation cannot be obtained (this would be impossible), the known distortion value of the signal generator was known, so it was possible to compensate for this distortion by subtracting it from the total figure that was obtained.

At this point, you may have a question regarding the amplification of the distortion which was already present in the signal generator. This distortion is amplified along with the rest of the signal, so if .5% distortion in the signal generator was amplified with the rest of the signal by a circuit with an amplification factor of five, then the distortion level would be five times higher than it was at its level within the generator. Remember, however, that the pure signal is also five times louder (or higher) in value, so the percentage of the distortion to the overall signal is still the same, and the subtraction of .5% from the total figure will still be correct.

Commercial radio stations must take distortion measurements on their entire system, as was mentioned earlier in this chapter. This includes the audio amplifier at the microphone input which is used by the announcer, any amplifiers which raise the level of the signal to be fed to the transmitter, and the modulation portion of the transmitter itself.

To perform these measurements, the output from the signal generator is fed to the microphone input at the microphone amplifier, which is generally part of the main control console. The signal is then amplified within the console and fed to the transmitter. Some means of demodulating the transmitting signal will be needed. There are a number of methods of doing this. One way would be to monitor the signal on a receiver, but if this were done, it would be difficult to determine the percentage of distortion within the receiver itself, thus interfering with the overall measurement.

A common practice which many radio stations utilize is to incorporate diode detectors. These may be in the form of a small loop of wire which is connected to the transmitter in such a way as to bring some of the rf down a piece of coaxial cable and into a circuit which is composed primarily of a solid-state diode. This circuit rectifies the rf signal and produces an audio output which is fed directly to the distortion meter. The diode detector is a passive unit. In other words, it does not amplify the signal it receives from the transmitter in any way and requires no external power to operate. It introduces almost no distortion into the circuit, and for most applications it is considered to be a device with no distortion of its own. Here, it can be seen that the entire transmitting system is being tested in a continuous loop for total distortion. The distortion introduced by the control console, the modulation section of the transmitter, and the transmitting stage is all in this continuous loop. Thus, the total distortion quantity of the entire signal is measured as it is broadcast.

Remember that it will be necessary to subtract the distortion content of the signal generator from the figure which is arrived at by this procedure. The signal generator is the only component in this loop which would not be included during normal broadcasting. It should be possible to determine the distortion which is introduced by the control board only by connecting the output of the control board directly to the distortion meter. The readings would be taken in the normal manner, with the distortion figure of the signal generator subtracted to obtain the overall distortion of the control console. To find the distortion introduced by the transmitter portion of the system, a test would be run as described earlier with all components in a continuous loop. The total distortion figure would indicate the transmitter distortion when the measurement of the control board is subtracted.

While the tests described thus far have been primarily for commercial applications, a distortion meter can be used for a home shop as well. Although it may be considered to be an unnecessary extravagance, it provides a simple and useful means of testing home projects once they are completed. Any type of audio circuit can easily be tested for distortion with this instrument. It may also be used to pinpoint problems in larger audio systems. If your home stereo system's audio quality is less than desirable, the distortion meter may be used to compare the distortion level in one channel to the same level in another channel. If one channel is producing a significantly higher amount of distortion than the other, this is a good indication that this one channel is at fault. If a public address (PA) system is being set up for a live musical concert or perhaps for some emergency purpose such as crowd control, the distortion meter can be used ahead of the scheduled time for the start of the event to check for any potential problems. This is accomplished by taking a distortion measurement at the time of the original purchase in order to ascertain what value is "normal." This level is then checked periodically and each and every time the equipment is expected to be put to use in order to determine any possible deterioration of vacuum tubes or other devices which control the quality and reliability of the output of the system or components.

It can be seen that the distortion meter is a useful device and can further enable you to realize the quality of home-built circuits as well as the operational characteristics of commercially manufactured devices.

NOISE MEASUREMENTS

A distortion meter is sometimes referred to as a *noise and distortion meter*, because the instrument can also provide its user with an accurate measurement of the noise level present in a device or system in addition to distortion. Of course, a portion of the distortion present in a device can be considered to be noise, which is simply the wide range of signals that are produced at a very low volume by components within the audio system. The finest amplifier will have a certain amount of noise at its output when the circuit is activated and no signal is being fed to its input. This is the internal noise of the circuitry.

In commercial applications, a specific noise level must be maintained by the equipment being used. This noise level will vary from circuit to circuit, but regardless of the amount of variation, it is always expressed in decibels (dB), or more appropriately, negative decibels. A noise level of −3 dB indicates that the level of noise is half as loud as the output of an audio amplifier. This noise level is taken with the amplifier producing an output of a specific level when a signal is fed to its input. This output level is used as a reference source. When the reference is set, the tone which is used as a driving signal for the distortion meter is completely removed. The distortion meter's function switch is thrown into the decibel position and lowered until an adequate reading is indicated on the meter. For example, if the range switch is in the −50 dB position and the meter indicates a reading of 3 dB, then the total noise level is equal to −53 dB.

The arrangement for measuring noise level is basically the same as the procedure described earlier in this chapter for measuring distortion. The similarity ends at the point where the reference is obtained by the incoming signal and the distortion meter is calibrated. In the case of taking a noise measurement, the main switch on the distortion meter is thrown to the decibel position instead of the distortion position. This is done after the driving signal from the audio generator is stopped.

Noise levels will vary with the position of the volume control on the audio amplifier under test. When the volume is high, the noise level will also be high. When the volume is lower, the noise level will be proportionally lower. Also, the level of internal noise will be proportionally higher or lower depending upon the level of the output signal at different volume settings. To illustrate this, if the volume were set at the halfway point, the noise level might be 50 dB below the level of the signal output from the amplifier. Alternately, if the volume control were set at the ¾ position, the noise level might be only 40 dB below the level of the output signal. In the latter position, the noise level actually produced is higher, but this value is always in reference to the value of the output signal.

All of the types of equipment which can be successfully measured for the level of distortion present can also be tested for noise level. This includes the example explained earlier for testing the distortion of a commercial radio station's entire system. These stations also test the noise level which is introduced into the system by the control console and the transmitter as well. Again, there will be a certain amount of noise which is due to the signal generator, and this will have to be taken into account when taking these measurements. Generally speaking, the procedure involves removing power from the signal generator so that no noise would be present from this source.

Testing a system or a device for noise level is an important indication of the quality of the unit or units. While the procedure for measuring noise and distortion described in this chapter may seem a bit complex at first, both are really very simple and only require a short period of time to become familiar with. After performing either of these tests once or twice, you should be able to take these measurements quickly and accurately.

Distortion measurements are normally taken at many different levels of audio input and output, as well as at various frequencies. This will aid in determining (in a way) the response of the apparatus under test to various input levels and frequencies. In this way, it will be possible to judge the low-frequency characteristics which lie just below the human hearing range, and the high-frequency characteristics which lie above the human hearing range and are transmitted through the audio system.

FREQUENCY RESPONSE

Many of the commercially manufactured noise and distortion meters can be used to determine the frequency response of various types of audio equipment. Frequency response is a very accurate name for a specific characteristic of a piece of audio equipment. This response is a measurement of how well various frequencies are amplified or reproduced by an audio amplifier or another type of audio equipment.

To illustrate, a frequency of 1000 Hz (when delivered to the input of an audio amplifier at a certain volume) will have an output from that amplifier which is also at a certain level. This frequency at its input level might serve as a reference for other measurements.

When a frequency of 3000 Hz is applied at the same level to the input of this same audio amplifier, the output will be at the same level as it was for the 1000-hertz tone if the amplifier has the same frequency response between 1000 and 3000 hertz. The chances are, however that the frequency response will be slightly different for the two frequencies, and the 3000-hertz signal may be slightly louder or softer at the output of the amplifier. If the output is slightly louder, then the audio amplifier has a better frequency response to a frequency of 3000 hertz than it does to a frequency of 1000 hertz or vice versa, depending upon the strength of the amplified signal.

Most noise and distortion meters are calibrated in decibels on one portion of the meter scale. In order to take frequency response measurements, it is first necessary to choose a reference signal. The 1000-hertz frequency is the one most often used. An input level must also be selected, one which can be monitored on the output meter of the signal generator if one is available. Alternately, if the audio amplifier or other piece of equipment to be tested has a level or VU meter, the input signal strength can be measured at this point.

The reference signal is fed to the audio equipment at a specific measured level. The output from the audio equipment is then connected to the noise and distortion meter. The signal output is adjusted to indicate a reading in the middle of the decibel scale. It makes no difference what this level is, as long as it is in the middle of the scale with the meter indicating an even number, such as 9 dB or 10 dB. This value will be a reference figure for a 1000-hertz tone, or whatever frequency is chosen for the test. Make certain that the amount of input signal is also known before proceeding.

To determine the audio equipment's response to different signals, choose a signal output from the generator which is higher or lower than the reference signal, and feed it to the input of the audio equipment at exactly the same level as the reference tone. In other words, make certain that the output meter on the generator reads the same for the new signal or that the VU meter is at the same level, whichever is applicable. It may be necessary at this point to adjust the volume control up or down on either the generator or the input to the audio equipment under test.

Once you have established that the new signal is at the same input level as the reference signal, observe the reading indicated on the decibel scale of the distortion meter. If the first signal drove the distortion meter to a value of 10 dB on the decibel scale and this new signal is reading 9 dB on the same scale, then the frequency response to the new signal is 1 dB below that of the reference signal. Another way of stating this same reading would be that the frequency response over the reference signal is −1 dB. If the reading were 12 dB on the decibel scale for the new signal, then this value would be 2 dB above the reference signal and would be expressed as +2 dB frequency response.

The frequency response is just as important as the noise and distortion characteristics of audio and stereo equipment. It also influences the quality of the component's output. A piece of audio equipment which has poor low frequency response will not properly broadcast the low-frequency sounds in the audio spectrum. When listening to music on a stereo system with poor low frequency response, the bass sound will not be heard very well. Likewise, a system with poor high-frequency response will sound very bassy because the higher frequencies are not being heard very well. Most manufacturers of audio equipment attempt to produce a flat frequency response throughout the range of the human ear. A flat frequency response is produced when the decibel rating from one end of the frequency range to the other does not vary greatly or at least cannot be detected by the unaided human ear. A frequency response which does not vary by more than about 3 decibels over a specific range is said to have a flat response, as the human ear cannot normally hear a difference this small. This is because a signal output which is 3 dB removed from a reference signal is being radiated with the equivalent of half the power output of the reference signal.

SUMMARY

Many home technicians have never taken distortion measurements or have had the need to do so. This is surprising, since any type of audio device can produce distortion when certain components begin to deteriorate. A regular distortion check of various audio equipments such as stereo receivers, tape players, etc., can be effective preventive maintenance. most people do not worry about distortion until it becomes very obvious and audible in the speakers. Chances are, however, that this distortion is the result of deterioration over a long period of time. With regular measurements of system distortion and frequency response, the overall performance of electronic equipment can be charted over a long period of time, and the deterioration can be observed and corrected before it becomes great enough to be noticeable.

Chapter 11

How to Buy Test Instruments

With the great multitude of test instruments available to the home technician on today's market, it is sometimes a difficult task to choose the instruments best suited to individual needs. How can you be sure that you are buying a quality instrument? Can you shop for bargains and still end up with equipment that is suitable? How do you tell a good instrument from a mediocre one? These and many more questions must be answered in order to arrive at a complement of instruments which represent an equitable balance between purchase price and performance.

From a pure quality standpoint, price is often a main determinant. Generally speaking, the higher priced instruments from companies which are best known for producing professional quality test equipment will be more accurate than those that are bargain priced. Simply put, it costs far more to produce a highly accurate instrument, due to the components involved, than to produce a less accurate instrument.

Let's take the ohmmeter or multimeter, for example. When reading resistance, the indicator depends upon internal resistors to serve as references. Some inexpensive multimeters will use resistors which have 20% tolerances. This means that a 1000-ohm resistor with a plus or minus 20% tolerance could have an actual value of from 800 to 1200 ohms. This further means that the final readings obtained on the meter or digital panel are accurate to within slightly less than plus or minus 20%. Additional deviations are incurred in the remainder of the circuitry which further reduce accuracy. Therefore, a multimeter which uses 20% resistors can be slightly more than 20% in error when taking any readings. This applies to ohms, volts, and current.

Using such a meter, a voltage reading of 10 volts dc might turn out to be actually 8 to 12 volts dc. the same deviation applies to current and resistance as well. Higher priced multimeters will use resistors with 10% tolerances. Here, a deviation of 10% of indicated value can exist. Moving up the scale, finer multimeters use resistors with 5% tolerance levels. A 5% deviation is certainly tolerable in most measurement situations. Some high-priced multimeters will use specially selected resistors which are rated to within plus or minus 1%. These instruments represent the ultimate in measurement accuracy. Their prices reflect their precision.

Obviously, resistors which are rated at a 1% tolerance are far more expensive than those with 20% values. Since many are used in multimeters, it is easy to see how the price can soar. High-tolerance components (often called precision components) can easily cost ten times those of standard tolerance. For example, a half-watt 5% resistor from one manufacturer's catalog sells for about 15¢ per unit. From the same manufacturer, a 1% tolerance resistor of the same value costs $1. This is nearly seven times the price of the previous component. Regardless of the electronic circuitry used, the tolerance level of the internal components will have a direct effect upon measurement accuracy and upon the overall price of the completed circuit.

Now, you must ask yourself if you really need a multimeter which is accurate to within plus or minus 1%. Few technicians really do. In servicing electronic equipment, the multimeter is used to run resistance, voltage, and current checks. Using the schematic drawings for the equipment being serviced, optimum readings are usually printed out at each circuit test point. If, for example, a certain circuit point should read 12 Vdc and you get a reading of 10 Vdc with a 20% tolerance multimeter, you can be pretty much assured that the circuit is operating within normal parameters. From a practical standpoint, even inexpensive multimeters have tolerance values of between five and ten percent. A 5% deviation in an optimum 12-volt reading would include an error factor of only .6 volt. At 10%, this would increase to 1.2 volts. When it becomes necessary to take fairly accurate or close readings, as long as you know the tolerance factor of your instrument, you are in a position to interpolate using the known percentage of variation which is inherent in your instrument.

It is even possible to make a deviation chart for your particular multimeter if you have access to one which offers very good accuracy. All that is necessary is to measure different levels of voltage, current, and resistance with both instruments and log the error factor on a graph. Here, the more accurate instrument is used as a point of reference. If you find that your less expensive instrument has an average deviation of plus 0.5 volts on one scale, then the instrument can be placarded as such. It will be necessary to run several comparison checks on each multimeter scale. By doing this extra work, you can end up with an inexpensive multimeter which contains the needed adjustment information to make it nearly as useful as a more accurate type.

Most instrument manufacturers include accuracy or stability specifications on their products in information sheets. By obtaining this data from several different companies, it is quite simple to make a practical comparison which you can then use to determine which instrument is most desirable for your needs.

Stability is an important factor in many test instruments, especially when applied to audio generators, frequency generators, and other devices which are designed to produce sine-wave outputs. Audio generators often present good stability in their upper frequency ranges but tend to drift off frequency at lower frequencies. I once experienced this situation during a proof-of-performance test I was conducting at a commercial radio station. During this testing, it was necessary to run checks on equipment performance while inserting different audio tones into the microphone channel of the control console. These frequencies ranged from 30 Hz to 7.5 kHz. At all frequencies above 400 Hz, the generator was quite stable. But below this level, there was a noticeable frequency drift which could easily be detected by the human ear. A noncommercial type of audio generator was being used for this test and proved to be as good as the more precise commercial models, except at the lower frequency range. The test was completed by borrowing a commercial audio generator to conduct the low-frequency portion of the test. A comparison was made at a later date between the two generators using sophisticated frequency meters. It was found that the inexpensive model compared favorably with the costly commercial unit in the high-frequency range, but was not competitive with the more expensive model below 400 Hz. From a practical standpoint, the inexpensive generator was and still is quite useful for 99% of all test purposes. So, with this type of normal usage, it would be a bit ridiculous to spend several times the cost of this unit for a more stable commercial model. The added cost could only be appreciated during 1% of its operational time. On the other hand, if the generator were to be used mainly for the purpose of conducting proof-of-performance tests, the more stable commercial instrument would be necessary.

The gist of all this is to have a general idea of the tasks for which any instrument you contemplate buying is to be used. Then, make your choice based upon this knowledge. Chances are, a very expensive instrument will not be needed in most cases.

Where do you buy test instruments? The choice is yours and can include a local hobby store or a major manufacturer of commercial quality devices. Chances are, you will buy some from one outlet and some from another.

For those readers who need commercial-quality instruments but really do not have the capital to purchase new equipment, industrial surplus catalogs will come in quite handy. Through these channels you will probably be able to buy used and fully operational commercial-grade instruments at a fraction of the original cost. Some of these instruments may even be brand new. Industrial surplus catalogs often offer new equipment from manufacturers who have updated their lines and been left with a fair number of the older models. They now dump these obsolete instruments, giving them in quantity to the industrial surplus catalog people, who pass savings along to you. It is quite possible that the surplus units are just as accurate as the newest models on the market today, but they probably won't look as modern and may not have as many operational features. There is a good chance that some of these added functions won't be needed anyway.

War surplus catalogs also offer many types of electronic test instruments such as frequency generators, frequency meters, audio generators, oscilloscopes, etc. Many of these products are left over from World War II, the Korean War, and even the Vietnam War. Depending upon the vintage of the equipment you purchase, many quite accurate instruments are available which may be selling for one cent on the dollar. Be extremely careful when purchasing war surplus equipment. Find out the condition, the availability of replacement parts, and general performance specifications. Some war surplus equipment may have been designed to operate from unusual ac power supplies. Standard household current is 115 or 230 volts ac at 60 hertz. Some surplus equipment may be designed to operate from the same voltage levels but at a frequency of 400 to 1200 hertz. These units will not operate directly from the ac line and must be converted by building new 60-hertz power supplies and incorporating them into these units.

Some war surplus oscilloscopes were designed for highly specialized purposes and may not be practical for standard electronic-servicing uses. While some may be modified, many are simply not worth the effort. Shop around, however, because you may find an excellent buy. For example, I was fortunate enough to obtain a Navy oscilloscope that had never been taken out of the original packing carton. The total price was $45.00, and this instrument offered all of the capabilities of modern commercial-quality scopes costing in excess of $2500. The surplus scope probably would have cost in excess of $5000 when new.

There are many good surplus buys in rf generators. Many boast very stable outputs and can tune a myriad of frequencies and harmonics of those frequencies. One, the BC-610, is quite popular even today, but make sure you get one with its operational manual intact. Without this manual, you really have no idea of what frequency you are tuning, because output frequency is determined by the setting of a digital dial that is not calibrated in hertz. To select a specific frequency, it is necessary to look in the manual to determine the proper dial setting. These devices are usually very inexpensive without the manuals. With the latter, they can cost three times as much. Most can be purchased, however, for less than $150.

One other important criterion to know when purchasing war surplus equipment is size and weight. About fifteen years ago, I was very excited to learn that I could obtain a 2000-watt linear amplifier through government surplus channels for less than $200, compared to about five times this amount for a newly manufactured unit which was sold through amateur radio outlets. Upon closer examination of the surplus buy, however, it was discovered that the complete unit weighed well over 1000 pounds. This would have made it completely unmanageable and therefore was not an attractive buy at any price. War surplus equipment tends to be very large. While 'he huge compartments may offer a wealth of salvageable parts; using a complete unit, especially in limited space situations which are often encountered in the home workshop and test bench, may be highly impractical.

Probably the best method of choosing a test instrument which will be suited for your needs is to use a similar instrument by the same manufacturer which may be owned by a friend. Alternately, an instrument which offers practically the same functions and tolerance levels as the one you intend to buy which is made by a different manufacturer will suffice. By doing this comparison, you will know far more about the performance and operation of the instrument before the actual purchase is made. Some companies may even allow you a thirty-day trial basis in which to become accustomed to the operation of an instrument. If it proves unsuitable, you can return it for a full refund.

We have already discussed the surplus markets, but there is another source of good, used equipment available in your own home town. Check with local electronic repair shops to see what they are using. Chances are, some of them may have instruments for sale. These may be in good working order, having only been replaced by more modern units. Some may be inoperational but will require only minor work to make them functional again. A visit to these repair shops is a good idea, even if you plan to purchase equipment directly from a manufacturer. Repair shops update equipment quite often and may even have the exact unit you hope to purchase on hand. With a little diplomacy, you can probably get the technician to comment on its performance and even allow you to use it for a few minutes on his bench.

Some manufacturers offer discounts if you purchase a combination of test instruments at one time. Since many test instruments are designed to be used in conjunction with each other as part of an overall system, buying a complete test bench setup usually means that the interconnecting cables are designed to mate with the connectors of all the equipment. This can save many hours of time in preparing specialized cords to interface different test units which do not share the same manufacturer.

When purchasing used equipment, look for signs of abuse or misuse. Most technicians take a great deal of pride in maintaining their instruments in the best-possible condition. Others may not be so careful. In electronic circuit work, especially in the field-testing end of it, instruments may receive a lot of inevitable rough handling. I have broken the faceplate on more than one multimeter during my servicing career, but it was always immediately replaced to avoid moisture contamination and further damage. An instrument with a damaged case, although apparently functioning properly, may be on its last legs. You might want to ask how long the equipment has been in this damaged state. If the answer is only a few days (and this can be verified), then there may be no problem. If the damage occurred years ago, beware. During the latter instance, it would be a good idea to remove the instrument from its case and closely examine the various components and wiring connections. An instrument which has been infiltrated with moisture reveals this quite readily upon visual inspection. Tarnish and even mold may cover the printed circuit board. You may even find traces of moisture in the form of water droplets within the circuit interior. Green, corroded leads are a sure giveaway (these leads will break under the slightest pressure). Also, look for signs of unprofessional repair work which may have been attempted by someone who knew little about service work. Unprofessional repair is far worse than no repair at all.

On the other hand, if the instrument is not operating properly but appears to be in a generally good state of repair when viewing its interior, this may be a worthwhile restoration project, especially if the device can be had at a very reasonable price.

What can go wrong with test instruments? Multimeters may be damaged by attempting to measure voltage and current values which lie beyond the range of the meter. This may result in a burned and open resistor, an open meter shunt, or damage to a few other relatively inexpensive components. You will definitely want to make sure that the meter movement is functioning properly, as this is the most expensive single component in most multimeters. A bent meter indicator (pointer or needle) is a sure sign that someone has tried to read a thousand-volt potential on the ten-volt scale, or in general, has used the meter improperly. This action could easily result in irreparable damage to the meter movement and possibly to other components as well. Attempting to read voltages which lie outside the range of a particular multimeter scale is quite common and usually does not result in catastrophic damage the first couple of times because of the built-in protective features of most multimeters. Continual abuse in this area, however, will certainly lead to problems with the instrument.

This brings up another point of consideration when contemplating the purchase of used equipment. Can you find out anything about the person or persons who used this equipment originally? Were they respected technicians or amateurish individuals with little or no respect for anything or anybody? If you find the latter to be true, it might be very wise to look elsewhere.

Test instruments are designed to take a lot of punishment. There are, of course, limits to what they will withstand, but most are not easily damaged beyond the point of practical repair unless this abuse has reached monumental proportions. I recall finding an old Simpson multimeter in the trash can at a radio station where I had recently been employed. Nontechnical station personnel had succeeded in damaging the device so badly that the needle indicator was broken at the midway point and the entire instrument had ceased to function. Simpson is a most respected name in the multimeter line, so with the permission of the station owner, this multimeter became the possession of the author. A quick inspection of the internal circuitry showed that a resistor and a shunt had been destroyed and a solder connection had also come loose. These problems were quickly repaired. But what do you do about the meter indicator? I did not wish to purchase a new meter because of its cost and the fact that I already owned a good Simpson multimeter. It was determined that the meter movement was fully operational, so an indicator needle was salvaged from another piece of equipment, and a drop of bonding cement was used to splice into the broken indicator.

This repair operation occurred more than fifteen years ago, and even today, I still use this much-abused multimeter almost daily. It is quite accurate but cannot be operated in the vertical position due to the slight excess weight created by the indicator splice. Other than this, it is a professional-quality instrument which was made operational for less than $3.00 and would cost well over $200 today. The moral to this story is to accept any free equipment you can get your hands on. If you can't repair it, you can certainly use some of the components as replacement parts for other instruments.

SUMMARY

There are many ways to go about buying test equipment. The devices you choose should be based upon your normal test instrument needs, the amount of money you have to put toward these devices, and future test endeavors. Most instruments will be purchased from the manufacturers who offer them in completed form, but a fair number of others will come from kit manufacturers, the surplus market, and through shops and individuals with used equipment in your area.

All of these resources have the potential for providing you with a complete test-instrument bench which will serve you for many years to come. Home technicians do not often replace test instruments on a regular basis. Chances are, the purchase you make today will still be serving you a decade from now. Make your decisions wisely and you will be rewarded for many years to come.

Appendix A

Semiconductor Letter Symbols

Letter symbols used in solid-state circuits are those proposed as standard for industry, or are special symbols not included as standard. Semiconductor symbols consist of a basic letter with subscripts, either alphabetical or numerical, or both, in accordance with the following rules:

1. A capital (upper case) letter designates external circuit parameters and components, large-signal device parameters, and maximum (peak), average (dc), or root-mean-square values of current, voltage, and power (I, V, P, etc.).

2. Instantaneous values of current, voltage, and power, which vary with time, and small-signal values are represented by the lower case (small) letter of the proper symbol (i, v, p, i_e, V_{eb}, etc.).

3. Dc values, instantaneous total values, and large-signal values, are indicated by capital subscripts (i_C, I_C, v_{EB}, V_{EB}, P_C, etc.).

4. Alternating component values are indicated by using lower case subscripts: note the examples i_c, I_c, v_{eb}, V_{eb}, P_c, p_c.

5. When it is necessary to distinguish between maximum, average, or root-mean-square values, maximum or average values may be represented by addition of a subscript m or av; examples are i_{cm}, I_{CM}, I_{cav}, i_{CAV}.

6. For electrical quantities, the first subscript designates the electrode at which the measurement is made.

7. For device parameters, the first subscript designates the element of the four-pole matrix; examples are I or i for input, O or o for output, F or f for forward transfer, and R or r for reverse transfer.

8. The second subscript normally designates the reference electrode.

9. Supply voltages are indicated by repeating the associated device electrode subscript, in which case, the reference terminal is then designated by the third subscript; note the cases V_{EE}, V_{CC}, V_{EEB}, V_{CCB}.

10. In devices having more than one terminal of the same type (say two bases), the terminal subscripts are modified by adding a number following the subscript and placed on the same line, for example, V_{B1-B2}.

11. In multiple-unit devices the terminal subscripts are modified by a number preceding the electrode subscript; note the example, V_{1B-2B}.

Semiconductor symbols change, and new symbols are developed to cover new devices as the art changes; an alphabetical list of the complex symbols is presented below for easy reference.

The list is divided into six sections. These sections are signal and rectifier diodes, zener diodes, thyristors and SCRs, transistors, unijunction transistors, and field-effect transistors.

Signal and Rectifier Diodes

PRV	Peak reverse voltage
I_o	Average rectifier forward current
I_r	Average reverse current
I_{surge}	Peak surge current
V_F	Average forward voltage drop
V_R	Dc blocking voltage

Zener Diodes

I_F	Forward current
I_Z	Zener current
I_{ZK}	Zener current near breakdown knee
I_{ZM}	Maximum dc zener current (limited by power dissipation)
I_{ZT}	Zener test current
V_f	Forward voltage
V_Z	Nominal zener voltage
Z_Z	Zener impedance

Z_{ZK}	Zener impedance near breakdown knee
Z_{ZT}	Zener impedance at zener test current
I_R	Reverse current
V_R	Reverse test voltage

Thyristors and SCRs

I_f	Forward current, rms value of forward anode current during the "on" state.
$I_{FM(pulse)}$	Repetitive pulse current. Repetitive peak forward anode current after application of gate signal for specified pulse conditions.
$I_{FM(surge)}$	Peak forward surge current. The maximum forward current having a single forward cycle in a 60 Hz single-phase resistive load system.
I_{FOM}	Peak forward blocking current, gate open. The maximum current through the thyristor when the device is in the "off" state for a stated anode-to-cathode voltage (anode positive) and junction temperature with the gate open.
I_{FXM}	Peak forward blocking current. Same as I_{FOM} except that the gate terminal is returned to the cathode through a stated impedance and/or bias voltage.
I_{GFM}	Peak forward gate current. The maximum instantaneous value of current which may flow between gate and cathode.
I_{GT}	Gate trigger current (continuous dc). The minimum dc gate current required to cause switching from the "off" state at a stated condition.
I_{HO}	Holding current. That value of forward anode current below which the controlled rectifier switches from the conducting state to the forward blocking condition with the gate open, at stated conditions.
I_{HX}	Holding current (gate connected). The value of forward anode current below which the controlled rectifier switches from the conducting state to the forward-blocking condition with the gate terminal returned to the cathode terminal through specified impedance and/or bias voltage.
$P_{F(AV)}$	Average forward power. Average value of power dissipation between anode and cathode.
P_{GFM}	Peak gate power. The maximum instantaneous value of gate power dissipation permitted.
I_{ROM}	Peak reverse-blocking current. The maximum current

through the thyristor when the device is in the reverse blocking state (anode negative) for a stated anode-to-cathode voltage and junction temperature with the gate open.

I_{RXM} — Peak reverse blocking current. Same as I_{ROM} except that the gate terminal is returned to the cathode through a stated impedance and/or bias voltage.

$P_{GF(AV)}$ — Average forward gate power. The value of maximum allowable gate dissipation averaged over a full cycle.

v_F — Forward "on" voltage. The voltage measured between anode and cathode during the "on" condition for specified conditions of anode and temperature.

$V_{F(on)}$ — Dynamic forward "on" voltage. The voltage measured between anode and cathode at a specified time after turn-on function has been initiated at stated conditions.

V_{FOM} — Peak forward-blocking voltage, gate open. The peak repetitive forward voltage which may be applied to the thyristor between anode and cathode (anode positive) with the gate open at stated conditions

V_{FXM} — Peak forward-blocking voltage. Same as V_{FOM} except that the gate terminal is returned to the cathode through a stated impedance and/or voltage.

V_{GFM} — Peak forward gate voltage. The maximum instantaneous voltage between the gate terminal and the cathode terminal resulting from the flow of forward gate current

V_{GRM} — Peak reverse gate voltage. The maximum instantaneous voltage which may be applied between the gate terminal and the cathode terminal when the junction between the gate region and the adjacent cathode region is reverse biased.

V_{GT} — Gate trigger voltage (continuous dc). The dc voltage between the gate and the cathode required to produce the dc gate trigger current.

$V_{ROM(rep)}$ — Peak reverse blocking voltage, gate open. The maximum allowable value of reverse voltage (repetitive or continuous dc) which can be applied between anode and cathode (anode negative) with the gate open for stated conditions.

V_{RXM} — Peak reverse-blocking voltage. Same as V_{ROM} except that the gate terminal is returned to the cathode through a stated impedance and/or bias voltage.

Transistors

A_G	Available gain
A_p	Power gain
A_I	Current gain
B or b	Base electrode
BV_{BCO}	Dc base-to-collector breakdown voltage, base reverse-biased with respect to collector, emitter open.
$^{BV}_{BEO}$	Dc base-to-emitter breakdown voltage, base reverse-biased with respect to emitter, collector open.
BV_{CBO}	Dc collector-to-base breakdown voltage, collector reverse-biased with respect to base, emitter open.
BV_{CEO}	Dc collector-to-emitter breakdown voltage, collector reverse-biased with respect to emitter, base open.
BV_{EBO}	Dc emitter-to-base breakdown voltage, emitter reverse-biased with respect to base, collector open.
BV_{ECO}	Dc emitter-to-collector breakdown voltage, emitter reverse-biased with respect to collector, base open.
C or c	Collector electrode
C_c	Collector junction capacitance
C_e	Emitter junction capacitance
C_{ib}, C_{ic} C_{ie}	Input capacitance for common base, collector, and emitter, respectively.
C_{ob}, C_{oc} C_{oe}	Output terminal capacitance, ac input open, for common base, collector and emitter, respectively.
D	Distortion
E or e	Emitter electrode
$f_{\alpha b}$, $f_{\alpha c}$, $f\alpha e$	Alpha cutoff frequency for common base, collector, and emitter, respectively.
f_{co}	Cutoff frequency
f_{max}	Maximum frequency of oscillation
GC (CB), GC (CC), GE (CE)	Grounded (or common) base, collector, and emitter, respectively.
G_b, G_c, G_e	Power gain for common base, collector, and emitter, respectively.
h	Hybrid parameter

h_{fe}, h_{fb}, h_{fc}	Small signal forward current transfer ratio, ac output shorted, common emitter, common base, common collector, respectively
h_{ib}	Small-signal input impedance, ac output shorted, common base.
h_{ob}	Small-signal output admittance, ac input open, common base.
I	Direct current (dc).
I_B, I_C, I_E	Dc current for base, collector, and emitter, respectively.
I_{CBO}	Dc collector current, collector reverse-biased with respect to base, emitter-to-base open.
I_{CES}	Dc collector current, collector reverse-biased with respect to emitter, base shorted to emitter.
I_{EBO}	Dc emitter current, emitter reverse-biased with respect to base, collector-to-base open.
NF	Noise figure
P_D	Total average power dissipation of all electrodes of a semiconductor device.
P_G	Power gain
P_{Go}	Overall power gain
P_{in}	Input power
P_{out}	Output power
$r'b$	Equivalent base resistance, high frequencies
T_j	Junction temperature
T_{stg}	Storage temperature
t_f	Fall time, from 90 percent to 10 percent of pulse (switching applications).
t_r	Rise time, from 10 percent to 90 percent pulse (switching applications).
t_s	Storage time (switching applications).
V_{BE}	Base-to-emitter dc voltage
V_{CE}	Collector-to-base dc voltage
V_{CE}	Collector-to-emitter dc voltage
V_{CEO}	Dc collector-to-emitter voltage with collector junction reverse-biased, zero base current.
V_{CER}	Similar to V_{CEO}, except with a resistor (of value R) between base and emitter.
V_{CES}	Similar to V_{CEO}, except with base shorted to emitter.

V_{CEV}	Dc collector-to-emitter voltage, used when only voltage bias is used.
V_{CEX}	Dc collector-to-emitter voltage, base-emitter back biased.
V_{EB}	Emitter-to-base dc voltage
V_{pt}	Punch-through voltage

Unijunction Transistors

I_E	Emitter current
I_{EO}	Emitter reverse current. Measured between emitter and base-two at a specified voltage, and base-one open-circuited.
I_p	Peak point emitter current. The maximum emitter current that can flow without allowing the UJT to go into the negative resistance region.
I_V	Valley point emitter. The current flowing in the emitter when the device is biased to the valley point.
r_{BB}	Interbase resistance. Resistance between base two and base one measured at a specified interbase voltage.
V_{B2B1}	Voltage between base two and base one. Positive at base two.
V_p	Peak point emitter voltage. The maximum voltage seen at the emitter before the UJT goes into the negative resistance region.
V_D	Forward voltage drop of the emitter junction
V_{EB1}	Emitter-to-base-one voltage.
$V_{EB1(SAT)}$	Emitter saturation voltage. Forward voltage drop from emitter to base one at a specified emitter current (larger than I_V) and specified interbase voltage.
V_v	Valley point emitter voltage. The voltage at which the valley point occurs with a specified V_{B2B1}.
V_{OB1}	Base one peak pulse voltage. The peak voltage measured across a resistor in series with base one when the UJT is operated as a relaxation oscillator in a specified circuit.
α_{rBB}	Interbase resistance temperature coefficient. Variation of resistance between B_2 and B_1 over the specified temperature range and measured at the specific interbase voltage and temperature with emitter open circuited.
$I_{B2(mod)}$	Interbase modulation current. B_2 current modulation due to firing. Measured at a specified interbase voltage, emitter and temperature.

299

Field-Effect Transistors

I_D — Drain current

I_{DGO} — Maximum leakage from drain to gate with source open

I_{DSS} — Drain current with gate connected to source

I_G — Gate current

I_{GSS} — Maximum gate current (leakage) with drain connected to source

$V_{(BR)DGO}$ — Drain to gate, source open

V_D — Dc drain voltage

$V_{(BR)DGS}$ — Drain to gate, source connected to drain

$V_{(BR)DS}$ — Drain to source, gate connection not specified

$V_{(BR)DSX}$ — Drain to source, gate biased to cutoff or beyond

$V_{(BR)GS}$ — Gate to source, drain connection not specified

$V_{(BR)GSS}$ — Gate to source, drain connected to source

$V_{(BR)GD}$ — Gate to drain, source connection not specified

$V_{(BR)GDS}$ — Gate to drain, source connected to drain

V_G — Dc gate voltage

$V_{G1S(OFF)}$ — Gate 1-source cutoff voltage (with gate 2 connected to source)

$V_{G2S(OFF)}$ — Gate 2-source cutoff voltage (with gate 1 connected to source)

$V_{GS(OFF)}$ — Cutoff voltage

Appendix B

Formulas

Ohm's Law for Dc Circuits

$$I = \frac{E}{R} = \frac{P}{E} = \sqrt{\frac{P}{R}}$$

$$R = \frac{E}{I} = \frac{P}{I^2} = \frac{E^2}{P}$$

$$E = IR = \frac{P}{I} = \sqrt{PR}$$

$$P = EI = \frac{E^2}{R} = I^2R$$

Resistors in Series

$$R_T = R_1 + R_2 + \ldots$$

Resistors in Parallel

Two resistors

$$R_T = \frac{R_1 R_2}{R_1 + R_2}$$

More than two

$$\frac{1}{R_T} = \frac{1}{R_1} + \frac{1}{R_2} + \frac{1}{R_3} + \ldots$$

RL Circuit Time Constant

$$\frac{L \text{ (in henrys)}}{R \text{ (in ohms)}} = t \text{ (in seconds), or}$$

$$\frac{L \text{ (in microhenrys)}}{R \text{ (in ohms)}} = t \text{ (in microseconds)}$$

RC Circuit Time Constant

R (ohms) × C (farads) = t (seconds)
R (megohms) × C (microfarads) = t (seconds)
R (ohms) × C (microfarads) = t (microseconds)
R (megohms) × C (micromicrofarads) = t (microseconds)

Capacitors in Series

Two capacitors

$$C_T = \frac{C_1 C_2}{C_1 + C_2}$$

More than two

$$\frac{1}{C_T} = \frac{1}{C_1} + \frac{1}{C_2} + \frac{1}{C_3} + \ldots$$

Capacitors in Parallel: $C_T = C_1 + C_2 + \ldots$

Capacitive Reactance: $X_C = \dfrac{1}{2\pi f C}$

Impedance in an RC Circuit (Series)

$$Z = \sqrt{R^2 + (X_C)^2}$$

Inductors in Series

$$L_T = L_1 + L_2 + \ldots \text{ (no coupling between coils)}$$

Inductors in Parallel

Two inductors

$$L_T = \frac{L_1 L_2}{L_1 + L_2} \quad \text{(no coupling between coils)}$$

More than two

$$\frac{1}{L_T} = \frac{1}{L_1} + \frac{1}{L_2} + \frac{1}{L_3} + \dots \text{(no coupling between coils)}$$

Inductive Reactance

$$X_L = 2\pi fL$$

Q of a Coil

$$Q = \frac{X_L}{R}$$

Impedance of an RL Circuit (Series)

$$Z = \sqrt{R^2 + (X_L)^2}$$

Impedance with R, C, and L in Series

$$Z = \sqrt{R^2 + (X_L - X_C)^2}$$

Parallel Circuit Impedance

$$Z = \frac{Z_1 Z_2}{Z_1 + Z_2}$$

Sinewave Voltage Relationships

Average value

$$E_{ave} = \frac{2}{\pi} \times E_{max} = 0.637 E_{max}$$

Effective or rms value

$$E_{eff} = \frac{E_{max}}{\sqrt{2}} = \frac{E_{max}}{1.414} = 0.707 E_{max}$$
$$= 1.11 \, E_{ave}$$

Maximum value

$$E_{max} = \sqrt{(E_{eff})} = 1.414 E_{eff}$$
$$= 1.57 E_{ave}$$

Voltage in an a-c circuit

$$E = IZ = \frac{P}{I \times P.F.}$$

303

Current in an a-c circuit

$$I = \frac{E}{Z} = \frac{P}{E \times P.F.}$$

Power in Ac Circuit

Apparent power: $P = EI$
True power: $P = EI \cos \theta = EI \times P.F.$
Power Factor

$$P.F. = \frac{P}{EI} = \cos \theta$$

$$\cos \theta = \frac{\text{true power}}{\text{apparent power}}$$

Transformers

Voltage relationship

$$\frac{E_p}{E_s} = \frac{N_p}{N_s} \text{ or } E_s = E_p \times \frac{N_s}{N_p}$$

Current relationship

$$\frac{I_p}{I_s} = \frac{N_s}{N_p}$$

Induced voltage

$$E_{eff} = 4.44 \times BAfN \times 10^{-8}$$

Turns ratio

$$\frac{N_p}{N_s} = \sqrt{\frac{Z_p}{Z_s}}$$

Secondary current

$$I_s = I_p \times \frac{N_p}{N_s}$$

Secondary voltage

$$E_s = E_p \times \frac{N_s}{N_p}$$

Three-Phase Voltage and Current Relationships

With wye connected windings

$$E_{line} = \sqrt{3}(E_{coil}) = 1.732 E_{coil}$$

$$I_{line} = I_{coil}$$

With delta connected windings

$$E_{line} = E_{coil}$$
$$I_{line} = 1.732 I_{coil}$$

With wye or delta connected winding

$$P_{coil} = E_{coil} I_{coil}$$
$$P_t = 3 P_{coil}$$
$$P_t = 1.732 E_{line} I_{line}$$

(To convert to true power multiply by cos θ)

Resonance

At resonance

$$X_L = X_C$$

Resonant frequency

$$F_o = \frac{1}{2\pi\sqrt{LC}}$$

Series resonance

$$Z \text{ (at any frequency)} = R + j(X_L - X_C)$$
$$Z \text{ (at resonance)} = R$$

Parallel resonance

$$Z_{max} \text{ (at resonance)} = \frac{X_L X_C}{R} = \frac{X_L^2}{R}$$

$$= QX_L = \frac{L}{CR}$$

Band width

$$\overset{\Delta}{=} \frac{F_o}{Q} = \frac{R}{2\pi L}$$

Tube Characteristics

Amplification factor

$$\mu = \frac{\Delta e_p}{\Delta e_g} \quad (i_p \text{ constant})$$
$$\mu = g_m r_p$$

305

Ac plate resistance

$$r_p = \frac{\Delta e_p}{\Delta i_p} \quad (e_g \text{ constant})$$

Grid-plate transconductance

$$g_m = \frac{\Delta i_p}{\Delta e_g} \quad (e_p \text{ constant})$$

Decibels

NOTE: Wherever the expression "log" appears without a subscript specifying the base, the logarithmic base is understood to be 10.

Power ratio

$$db = 10 \log \frac{P_2}{P_1}$$

Current and voltage ratio

$$db = 20 \log \frac{I_2 \sqrt{R_2}}{I_1 \sqrt{R_1}}$$

$$db = 20 \log \frac{E_2 \sqrt{R_1}}{E_1 \sqrt{R_2}}$$

NOTE: When R_1 and R_2 are equal they may be omitted from the formula. When reference level is one milliwatt

$$dbm = 10 \log \frac{P}{0.001} \quad (\text{when P is in watts})$$

Synchronous Speed of Motor

$$r.p.m. = \frac{120 \times \text{frequency}}{\text{number of poles}}$$

Index

components of, 170
electron gun use in, 173
electrostatic deflection in, 176-179
electrostatic vs. electromagnetic, 171
focusing action in, 174
phosphor screen of, 171
waveform display in, 179
Centigrade scale, 9
centimeter-gram-second (cgs) system, 4
ceramic capacitors, 43
chop mode (see dual-trace oscilloscope)
circuit boards, 240
circuit testing, 70-92
failure analysis in, 82
listing probable faulty functions in, 76
localizing faulty function in, 78
localizing trouble to circuits in, 81
symptom elaboration, 72-76
symptom recognition in, 71
use of voltmeter in, 89
coefficient of coupling, 31
coils, 33, 101, 303
collector current, 47
complex waves, 199
components, mounting of, 244
conductance, 21-23
conductors, 98
core material in, 102
left-hand rule for, 96, 97
magnetic field distortion by, 103
right-hand rule of, 104
strength of, 101
continuity tester, homebuilt, 248
convergent-divergent path, 81-84
Coulomb's law, 25
counter electromotive force (CEMF), 28, 33
coupling
coefficient of, 31
inductive, 32
crossover, 208
current, 18-21
flow of (see conductors, electromagnetism), 98-101
magnitude of, 19
multimeter measurement of, 141-144
reversal of, 104
three-phase voltage relationship to, 304

D

D'Arsonval-type meter movement, 105-108
dc ammeter, 108-111, 108
dc blocking voltage, 52
dc voltage, 18
multimeter measurement of, 138-141
dc voltmeter, 112-114
decibels, equations for, 306
deflection factor, 178
delay line (see dual-trace oscilloscope)
depletion MOSFET, 49
device dissipation, 46
dielectrics, 26
diodes, 49-55
direct current (dc), 19
directed drift, 19
distortion, 188
allowable guidelines for, 275
frequency response and, 282
measurement of, 274-284
noise measurement and, 280
using signal generator to measure, 277-280
distortion meter, 275
divergent path, 81, 83
dual in-line package vs. flat package, 244
dual-trace oscilloscope, 209
accessories for, 226
controls on, 211-222
delay line in, 224
operation of, 223
probe tips for, 227
time-base generator in, 224
unblanking of, 225
dummy loads (see wattmeter)
dyne, 10

E

electricity use, metering of, 126
electrodynamometer type voltmeter, 121
electrolytic capacitors, 43, 44
electromagnetic spectrum, 8
electromagnetism, 96-105
electromotive force, 18
counter (CEMF), 28, 33
electron beams, 170
electron gun, 170-173
electron movement, 20
electronic test instruments, 146-167

Other Bestsellers of Related Interest

POWER CONTROL WITH SOLID-STATE DEVICES
—Irving M. Gottlieb
  You'll find yourself turning to this book again and again as a quick reference *and* as a ready source of circuit ideas. Author Irving Gottlieb, a professional engineer involved in power engineering and electronic circuit design, examines both basic and state-of-the-art power control devices. 384 pages, 236 illustrations. Book No. 2795, $23.95 paperback, $29.95 hardcover

ELECTRONIC CONVERSIONS, SYMBOLS AND FORMULAS
—2nd Edition—Rufus P. Turner and Stan Gibilisco
  This revised and updated edition supplies all the formulas, symbols, tables, and conversion factors commonly used in electronics. Exceptionally easy to use, the material is organized by subject matter. Its format is ideal and you can save time by directly accessing specific information. Topics cover only the most-needed facts about the most often used conversions, symbols, formulas, and tables. Basic mathematics are presented first, and follow-on formulas build from that base. 280 pages, 94 illustrations. Book No. 2865, $14.95 paperback, $21.95 hardcover

ELEMENTARY ELECTRICITY AND ELECTRONICS
—COMPONENT BY COMPONENT—Mannie Horowitz
  Here's a comprehensive overview of fundamental electronics principles using specific components to illustrate and explain each concept. This approach is particularity effective because it allows you to easily learn component symbols and how to read schematic diagrams as you master theoretical concepts. You'll be led, step-by-step, through electronic components and their circuit applications. Horowitz has also included an introduction to digital electronics. 350 pages, 231 illustrations. Book No. 2753, $16.95 paperback, $23.95 hardcover

ELECTRONICS DISPLAY DEVICES—Richard A. Perez
  The first complete study of its kind, this exceptional edition reviews all display technologies on the market today. Discover what's on the horizon of developing electronic display devices—from CRTs, LEDs, and LCDs to ELDs, plasma, and VFDs. This exciting look at electronic displays explains everything you need to know. The author describes what electronics displaydevices are, how they work, and how you can best use them. 416 pages, 192 illustrations. Book No. 2957, $39.95 hardcover only

DIGITAL ELECTRONICS TROUBLESHOOTING—2nd Edition
—Joseph J. Carr

Now, the Electronics Book Club brings you a brand new, completely updated, and expanded edition of this classic guide to digital electronics troubleshooting. It converts not only the basics of digital circuitry found in the first edition, it also provides details on several forms of clock circuits, including up-to-the-minute coverage of microprocessors used in today's cassette players, VCRs, TV sets, auto fuel and ignition systems, and many other consumer products. 428 pages, 324 illustrations. Book No. 2750, $17.95 paperback only

Prices Subject to Change Without Notice.

Look for These and Other TAB Books at Your Local Bookstore

To Order Call Toll Free 1-800-822-8158
(in PA, AK, and Canada call 717-794-2191)
or write to TAB BOOKS, Blue Ridge Summit, PA 17294-0840.

Title	Product No.	Quantity	Price

☐ Check or money order made payable to TAB BOOKS

Charge my ☐ VISA ☐ MasterCard ☐ American Express

Acct. No. _____ Exp. _____

Signature: _____

Name: _____

Address: _____

City: _____

State: _____ Zip: _____

TAB BOOKS catalog free with purchase; otherwise send $1.00 in check or money order and receive $1.00 credit on your next purchase.

Orders outside U.S. must pay with international money order in U.S. dollars.

TAB Guarantee: If for any reason you are not satisfied with the book(s) you order, simply return it (them) within 15 days and receive a full refund.   BC

Subtotal $ _____

Postage and Handling
($3.00 in U.S.,
$5.00 outside U.S.) $ _____

Add applicable state
and local sales tax $ _____

TOTAL $ _____